Tributes
Volume 47

Festschrift for Martin Purvis
An Information Science "Renaissance Man"

Tributes Series Editor
Dov Gabbay

dov.gabbay@kcl.ac.uk

Festschrift for Martin Purvis
An Information Science "Renaissance Man"

edited by

Mariusz Nowostawski

Holger Regenbrecht

ISBN 978-1-84890-410-1

College Publications
Scientific Director: Dov Gabbay
Managing Director: Jane Spurr

http://www.collegepublications.co.uk

Cover design by Laraine Welch

CONTENTS

ARTICLES

LETTERS

Foreword

MARIUSZ NOWOSTAWSKI
National Technical University of Norway

HOLGER REGENBRECHT
University of Otago, Dunedin, New Zealand

As guest editors we are thrilled to present this festschrift in honour of Emeritus Professor Martin K. Purvis. We follow the academic custom of honouring important milestones in a colleague's career with such a collected volume of scholarly articles and reflective notes by current and former colleagues of the honoured individual.

Often, a festschrift is compiled to celebrate the retirement of an outstanding senior researcher or teacher. However, Prof. Purvis rejects the notion of retirement and is staying active in academic life, so we have taken the liberty to put together a festschrift without waiting for a farewell event to present this to him!

Martin Purvis received his B.S. in Physics from Yale University in 1967 and his M.S. and Ph.D. in Physics from the University of Massachusetts in 1971 and 1974, respectively. He also graduated with a Master of Fine Arts in Film from Columbia University in 1980.

He was an Assistant Professor of Physics at Pars College, Tehran, Iran (1976–77) and Columbia University, New York, NY, USA (1977–1981). He worked as a Senior Research Scientist with the LTV Aerospace and Defence Company in Dallas, TX, USA (1982–1984), as a Senior Operating Systems Specialist at the University of Texas, Austin, TX, USA (1984–1989), and was a member of technical staff at Microelectronics and Computer Technology Corporation (MCC) in Austin (1990–1992).

Since 1992 he is with the Department of Information Science at the University of Otago, Aotearoa, New Zealand and made his way up the academic ranks from Senior Lecturer (1992–1999), Associate Professor (2000–2002) to Full Professor (2003–2015). Since then he is an Emeritus Professor with the department.

Amongst his many academic achievements, he was the initiator and director of the software engineering and telecommunications programmes in Applied Sciences, Director of the Software Engineering and Collaborative Modelling Laboratory and the Knowledge, Intelligence, and Web Informatics Laboratory, and the Coordinator of the Connectionist-Based Information Systems Research Theme at the University of Otago.

Prof. Purvis published close to 300 academic articles and has about 4,000 citations to his work. He is an internationally renowned expert and leading figure especially in the areas of software engineering, including innovative teaching methods and multi-agent systems modelling, with a particular focus on norms in social agent societies. Through his intensive and extensive inter- and intra-disciplinary collaborations he contributed significantly to other foreign and adjacent research areas such as neural networks, fuzzy systems, spatial information modelling, and sustainability modelling and analysis.

In this festschrift we have the pleasure to present a selection of academic articles dedicated to Prof. Purvis which address the breadth and depth of Martin's interests and expertise. They are written by former and current colleagues and postgraduate students of his. In addition, six people contributed personal notes for Martin which are presented at the end of this festschrift.

Honoured to be Prof. Purvis' colleagues.

Mariusz & Holger

ARTICLES

Human Idiosyncrasies and the Emergence of Science

Carl Stuart Leichter PhD
Norwegian University of Science and Technology, NTNU
dr.carl.leichter@gmail.com

Abstract

Scientific knowledge is the result of interactions between human institutions and the scientific domains they study. Besides the theories and experimental methods reported in research publications, these interactions also include interpersonal relationships among scientists, conformance with[1] academic institutional norms and general engagement with society. Given this complex and interactive model of scientific inquiry, then the scientific theories produced by science are 'emergent properties' of interactive systems. As such, these theories should exhibit the general characteristics of emergent properties; in particular the theories should be robust and autonomous from their constituent (human) components. Additionally, the institutions of science are highly non-linear, dynamic systems; in extreme circumstances their institutional evolutionary processes will exhibit the unstable, far-from equilibrium dynamics described by Ilya Prigogine's theory dissipative structures. Specifically, in times of crisis, the destabilized institutions of science will be highly sensitive to the minute perturbations of unscientific influences. In this paper, we will combine the principles of emergence, the theory of dissipative structures and the historical perspective of Thomas Kuhn's structure of scientific revolutions into a model for "dissipative structure emergence of scientific revolutions". We will apply this model to a case history of emergence in science by recounting some of the key

---

[1]Rebellion against?

events in the early 20th Century formulation of quantum mechanics. We will find that the robust and autonomous science that emerged was shaped by the minute perturbations of human idiosyncrasies.

Part I
Introduction

1 Overview

The origins of scientific inquiry stretch back thousands of years. The Babylonians used astronomical observations to predict solar eclipses [24], while in ancient Greece, Archimedes developed an algorithm for computing π and Pythagoras derived his eponymous theorem. But if we choose the founding of the Royal Society[2] as a starting point, then *modern* science[3] as we know it has only emerged over the last few hundred years. Besides its interactions with the world it is studying, science is also a system of humans interacting with each other. Therefore scientific knowledge is an *emergent property* of these human-centric system interactions. Additionally, like all systems, the institutions of science are subject to the principles of system dynamics. So the emergence of scientific knowledge is an evolutionary process which should exhibit non-linear, dynamic behavior.

1.1 Article Organization

Part II will present the theoretical and conceptual background for this paper: section 2 briefly discusses the concept of emergence and section 3 presents a few relevant aspects of nonlinear system dynamics, including the theory of dissipative structures. These concepts are com-

---

[2]The Royal Society of London for Improving Natural Knowledge est AD1660

[3]In this article the word "science" will always refer to the human institutions and undertakings of science, including the theories it creates; it will not refer to the natural world that is studied by science.

bined in section 4 and applied to Kuhn's model for paradigm change in science. The resulting model is demonstrated in part III, where we present "A Case Study of Emergence in Science: The Formulation of Quantum Mechanics".

Part II
Background

2 Emergence

Systems of interacting components often exhibit emergent properties which differ from the properties of the system's constituent components. The properties that emerge are the result of an "ensemble behavior" that arises when the system components interact with each other and their environment[7]. So by definition, emergent system properties cannot be predicted through an internal analysis of the individual system components[2]. In this article we will consider two types of emergence: "weak" and "strong" [3]. In the weak case, the system properties that emerge are unexpected; but they can be predicted through the analysis of ensemble behavior and/or component interactions. On the other hand, strongly emergent properties cannot be predicted, not even in principle. In both cases, the emergent properties are autonomous from the individual system components. This means emergent properties have sufficient robustness to still emerge, even when there are changes to the components from which they arise [6, p.2].

2.1 Weak Emergence

The term "weak emergence" refers to unexpected high level domain phenomena, or properties, that are derivable from the low level domain's constituents/processes. Statistical thermodynamics and the Central Limit Theorem are examples weak emergence. The CLT demonstrates that a normal/Gaussian distribution is derivable from a mixture of non-Gaussian random variables [14, p.157]. In the case of statistical thermodynamics: if we are given the Boltzmann constant, then by analyzing the quantum energies in an ensemble of atoms we can derive the fundamental temperature [9, p.41] and the Ideal Gas Law [9, p.76].

2.2 Strong Emergence

Strongly emergent properties cannot be derived,not even in principle, from the underlying constituents/processes. For example, the ideal gas law shown in equation 1 [5, p.476] can be derived *if we are given the universal gas constant*[4], *R*. This equation is a consolidation of the gas laws shown in figure 1; it demonstrates the relationships between the pressure (P) exerted by a collection of gas particles in moles (n), with thermal energy temperature (T) that are contained within a container of volume (V).

$$PV = nRT \tag{1}$$

But the value of the constant R itself cannot be analytically derived, it can only be empirically determined; because its value only emerges when a critical mass of gas particles are physically interacting with each other:

> "[The exact value of some fundamental constants] is a collective effect......R is known to an accuracy of one part
> in one million, yet it acquires huge errors in gas samples

---

[4]In some formulations, the ideal gas law is given in terms of the number of gas particles N and the Boltzmann constant k; those have $PV = NkT$. In the formulation used in this text, R is the product of Avogadro's number N_A with the Boltzmann constant: $R = N_A k$. The universal gas constant and the Boltzmann constant are the same strongly emergent physical property, they just have different units.

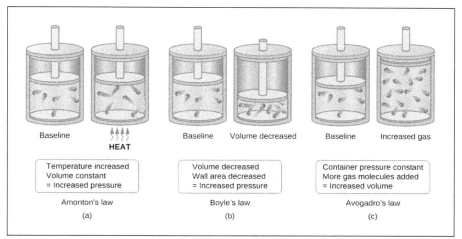

Figure 1: Gas Law Components.

that are too small, and ceases to be measurable at all, at the level of a few atoms" [12, p.20].

2.3   Autonomy in Emergence

The principle of autonomy means that an emergent property is not dependent upon its constituent components; so systems that have similar or identical system structures, but have different constituent components, may give rise to the same emergent property. Additionally, changes to the constituent components in a system may have no effect on the system's emergent properties. For example, in the case of weak emergence: the Central Limit Theorem holds even if the constituent components of the mixture have been changed. In the case of strong emergence: Van Der Waals' equation demonstrates the autonomy of the universal gas constant.

The Ideal Gas Law has two unrealistic assumptions: (i) gas particles have fully elastic collisions and (ii) they occupy zero volume in space. Van Der Waals' equation shown in eq (2) [13, p.6–64] corrects these assumptions by introducing an inelastic collision correction fac-

9

tor "a" and a particle volume factor "b". While factors "a" and "b" have a wide range of values that reflect the different Coulombic forces and sizes of different gases, the universal gas constant itself has the same value in all cases. It doesn't matter if the gas is composed of mono-atomic helium, diatomic oxygen, hydrocarbon gases such as CH_4 or even mixtures of all of the above: the universal gas constant remains the same. N.B. In cases where the gas is near its critical temperature, then the Van Der Waal's equation must use additional correction factors; but even in those cases, the universal gas constant remains.....constant [9, p.290].

$$(P + \frac{an^2}{V^2})(V - nb) = nRT \tag{2}$$

3 Dissipative Structures

Ilya Prigogine's Nobel Prize winning "Theory of Dissipative Structures" is a framework that explains how non-linear, dynamic system components can spontaneously self-organize[5] into complex structures and exhibit emergent properties which differ from their constituent components [19]. This theory describes how a non-linear system that had been in a stable, dynamic equilibrium with its environment can be driven into an unstable, far from equilibrium regime of chaotic behavior which can lead to the emergence of a new system order: the dissipative structure.

3.1 Dynamic Systems in Stable Equilibrium With Their Environment

The theory of dissipative structures is applicable across a wide variety of problem domains, because the principles of non-linear dynamics are general in nature: they can be applied in physics, chemistry, biology, sociology, economics or in any domain where the system under study can be described by using non-linear mathematics [21] . For example,

---

[5]His theory also explains that self organization is compatible with the second law of thermodynamics.

the Belousov–Zhabotinsky reaction is considered a classic case of an excitable, non-linear, oscillatory chemical system [18]. A schematic representation [19, p.168] of its behavior is shown in figure 2. It illustrates the different phases for a non-linear, dynamic system it is driven from its simplest steady state through increasing complex states, into chaotic behavior and then beyond.

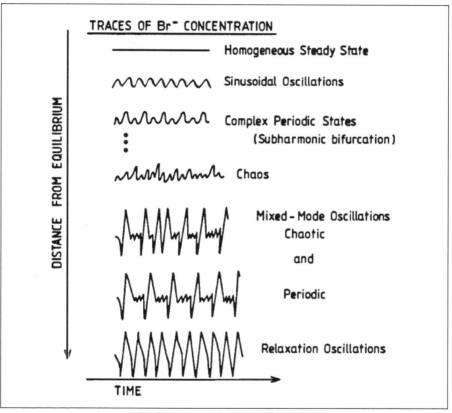

Figure 2: Schematic of Br- Oscillations in Belousov–Zhabotinsky reaction.

3.2 Dynamic Systems Driven To Unstable,

Far From Equilibrium States

Figure 2 serves to illustrate non-linear system evolution, when changing conditions *drive* the system out of a stable, dynamic equilibrium state. This may be due to environmental changes (eg: changes in available free energy) or because of internal changes to the system (eg: system decay). As the system is driven further and further from equilibrium, then it will gradually become more and more unstable and chaotic. After it has been driven into a far from equilibrium state, the system may spontaneously disintegrate or it may self-organize into a new, dynamic stable state that is capable of dissipating the perturbances (eg: increased thermal energy gradients).

3.3 Instability and Sensitivity: Perturbations and Bifurcations

3.3.1 Instability and Perturbation

When nonlinear systems are driven into a far from equilibrium state, they enter a region of highly divergent trajectories and so they are hypersensitive to minute perturbations. This sensitivity is popularly known as 'the butterfly effect'. It simply means that a very minute perturbation will be greatly amplified by the trajectory divergences. The difference between system stability in equilibrium versus sensitivity in far-from-equilibrium states, is illustrated by the thought experiment shown in figure 3. The pyramid (A) on the left is in a stable equilibrium state: resting on its base. If a person were to gently blow air on it, then it would remain as it is. But if the pyramid were rotated clockwise 180 degrees, so it's perfectly balanced on a vertex as in (B), then it has been driven into an unstable state. In this case, a slight perturbation (such as blowing a puff of air at it) would cause it to fall over into the new stable state (C). If the puff of air came from a different direction, then it would have fallen on a different side. In this particular example, from the unstable state (B), there are four different stable states the pyramid can fall into: the left side, right side, near side or the far side[6].

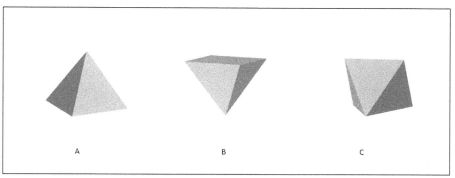

Figure 3: Equilibrium of a Pyramid.

3.3.2 Bifurcation Points

When a system is driven to a far from equilibrium condition, its trajectory may reach a bifurcation point. This is a point that can lead to several different possible new stable states. If the system is left undisturbed by external perturbations, then the "choice" of stable state is a random process. However, the slightest force applied to the system at its bifurcation point can determine the "choice" made. For example, in the case of the pyramid illustration above, there were four new stable states available and a slight puff of air could determine which one is entered.

The diagram in fig. 4 from [23] shows a mathematical demonstration of bifurcation. This is a plot of descending values of the dampening factor "a" against the maximum solution values for the fourth order differential equation (3). As dampening is reduced, the system becomes less and less stable. For illustration purposes, assume that this equation represents a non-linear, dynamic system. The specific instantiation of the system is irrelevant: it could be a system of chemical reactions or a mechanical system, or an electronic system, etc. As

---

[6]www.thefarside.com

13

the dampening factor decreases from 2.1 to 2.08, the max solutions for the system are stable and determined by initial conditions. But as the dampening factor is decreased to ~ 2.07, then the max solution trajectory reaches a point of instability. This is the bifurcation point identified by red arrow "1", where the trajectory branches out into two max solution regions. When this system reaches the bifurcation point, it is analogous to the pyramid (B) in fig. 3; here the trajectory path selection can be determined by a minute fluctuation at the bifurcation point: such as a tiny puff of air generated by the proverbial butterfly wing[7]. The system sensitivity to fluctuations is also illustrated by the chaotic proliferation of solutions in the region encountered when the dampening factor is reduced below 2.06, identified by the second red arrow.

$$\dddot{x} + a\,\ddot{x} \pm \ddot{x}^2 + x = 0 \tag{3}$$

3.4 Dissipative Structures

A dissipative structure may form when a dynamic system is driven into a far from equilibrium state. Figure 5 shows the difference between laminar and turbulent flow in fluid dynamics [15]. The laminar flow regime is a stable dynamic state. But when the flow rate exceeds a critical threshold[8], then laminar flow breaks down and the system enters a turbulent flow regime [20]. The vortices that form within the turbulent flow are classic examples of dissipative structures: they are self-organized structures that have emerged from the chaos of turbulent flow. Just before a vortex is formed, the laminar flow has reached a bifurcation point and subsequent vortex formation is extremely sensitive to any localized perturbations. Recalling the example in figure 3; it illustrates that there could be several different potential dissi-

---

[7]If a non-linear deterministic process has been driven to a perfectly balanced bifurcation point, then its trajectory will be influenced by the random fluctuations of quantum phenomena. This will effectively convert the deterministic process into a stochastic one.

[8]Determined by the Reynolds number.

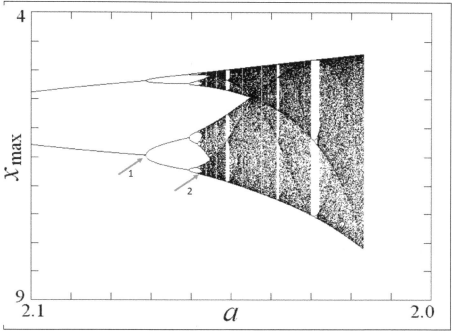

Figure 4: Bifurcation Diagram For Eq. 3

pative structures and the one(s) that emerge(s) can be influenced or determined by a minute perturbation.

4 The Institution of Science:
A Dynamic System of Interacting Humans

If we assume that the essential proprieties of reality (eg: speed of light, Planks constant, π, etc.) are constant, then changes in science are not caused by a changes in reality; instead they emerge from a restructuring of our models of reality. Because of the complex nature of modern scientific theories, the evolution of these models are not only influenced by empirical data, they are also influenced by interactions amongst individual scientists and institutions. Therefore we should

15

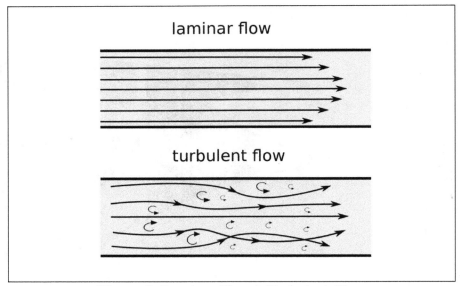

Figure 5: Laminar Versus Turbulent Flow

consider the modern institution of science to be a complex, interactive, dynamic system and like any other system: it can be in a dynamic, stable equilibrium state or it can be driven to an unstable, far from equilibrium state. If science is driven to a far from equilibrium state, then a new stable state (dissipative structure) can emerge. Just as with the non-linear dynamics of dissipative structures discussed in Section 3, this emergence can be influenced or determined by minute perturbations.

4.1 Science in an Equilibrium and then Far From Equilibrium State

Kuhn's classic "The Structure of Scientific Revolutions" is an in depth analysis of the behavior of science as a human institution [10]. It introduces the concept of a scientific "paradigm" to partition scientific activity into two phases: *ordinary science* and *extraordinary science*. During the ordinary science phase, the accepted paradigm is a theoretical model of how the universe works and it governs scientific inquiry. The extraordinary science phase is triggered, when new data emerges that falsifies the accepted paradigm.

4.1.1 Ordinary Science

The ordinary science phase refers to the exploration and elaboration of the currently accepted paradigm. It has been accepted, because it gives the best[9] explanation for the data in hand. For example: a scientist who accepted the Newtonian paradigm would not attempt to demonstrate perpetual motion. But they might attempt to demonstrate variations in the speed of light due to aether drift[10]. During 'ordinary' science, the accepted paradigm sets parameters for further inquiry and it may be extended into many different sub-domains.

4.1.2 Extraordinary Science

Kuhn asserts that fundamental scientific change occurs during the phase he identifies as "extraordinary science", which is triggered when an experiment falsifies a core principle of the accepted paradigm. For example, when the Michelson-Morley experiment demonstrated that the speed of light was constant in a vacuum, it falsified a core principle of Newtonian Mechanics. This falsification triggered a phase of extraordinary science that eventually led to a new paradigm: Relativistic Mechanics.

If Prigogine's dissipative structures is combined with Kuhn's model of scientific revolutions, then ordinary science takes place when the system of science is in an equilibrium state and extraordinary science occurs when science has been driven into a far from equilibrium state. When Alvin Toffler wrote his forward for "Order Out of Chaos", he

---

[9]Generally, the simplest and easiest to use theoretical model. This is why wave mechanics are used for QM, instead of matrix mechanics. Of course, it is far more complex than that; but that is another topic beyond our scope.

[10]Michelson and Morley 1887

describes the scientific revolution caused by Newtonian Mechanics in the 17th Century [19, p.xxvi] in these terms:

> [The] Newtonian knowledge system [is] itself, a "cultural dissipative structure" born of social fluctuation...

Recounting the circumstances surrounding the Newtonian scientific revolution of the 17th century is beyond the scope of this article. But it was also an example of the robust emergence of science from highly idiosyncratic human interactions: Issac Newton's relationships with his fellow scientists were extremely non-linear and often chaotic.

4.1.3 The Process Driving Science Into a Far From Equilibrium State: Aggregation of Inquiry

When an ordinary science experiment yields an anomalous result that threatens the accepted paradigm, it is initially known only to its discoverers. Eventually they publish their findings and other scientists, who have accepted the same paradigm, become aware of the anomaly. Some of these other scientists may begin their own inquiries into the anomaly and then publish their findings, which will get the attention of even more scientists...and so on. This continues until the leading researchers in a scientific community have "aggregated" around the inconsistency in the paradigm. At this point, the system of science has been driven into the far from equilibrium state that Kuhn has identified as 'extraordinary science'. A relatively contemporary example of this aggregation process occurred after the discovery of high temperature superconductors in the late 1980s[11].

4.1.4 Science in a Far From Equilibrium State

As seen in fig.4 on pg. 11, when a non-linear, dynamic system is driven into an unstable far from equilibrium state, it may reach a bifurcation point that branches out into several possible new stable states. Like

---

[11]https://ieeecsc.org/event/12th-international-conference-materials-and-mechanisms-superconductivity-high-temperature

any other dynamic system, when the human system of science has been driven out of equilibrium to a bifurcation point, it may also have several possible stable states (new paradigms) to choose from. In the case of Quantum Mechanics there were two new paradigms available: the particle perspective of Heisenberg's Matrix Mechanics and the wave perspective of Schrödinger's Wave Mechanics.

Part III
A Case Study of Emergence in Science: The Formulation of Quantum Mechanics

5 The Aggregation Of Scientific Inquiry Around an Anomaly

Just prior to the beginning of the twentieth century, anomalous phenomena in physics were observed that falsified the classical mechanics' predictions for black body radiation. At higher frequencies (shorter wavelengths), the radiation emissions observed in experiments did not agree with classical theory. The differences between the blue and black traces in fig. 6 illustrate this issue. The blue line represents the observed radiation emitted by black bodies that have been heated to 5000 K, while the black line represents the emissions predicted by classical theory.

19

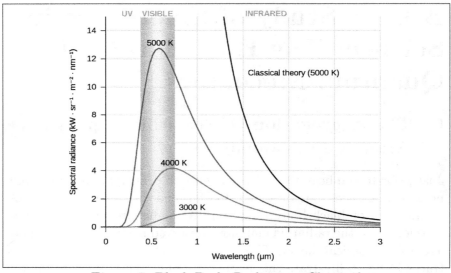

Figure 6: Black Body Radiation: Classical
Theory vs Empirical Observation.

This anomaly would later be referred to as the "ultra-violet catastrophe" and Max Planck partially resolved it in 1900. Plank hypothesized that light was not the continuous phenomenon expected by classical mechanics; instead it was composed of discrete energy packets that he called "quanta". In 1905, Einstein expanded on Plank's work, by demonstrating[12] the photoelectric effect. From there, the aggregation that would lead to the quantum revolution accelerated. By 1911, after the first European conference on quanta, Henri Poincare observed (emphasis added):

> ...[quantum theory would] be the greatest and most radical *revolution* in natural philosophy since the time of Newton..." [8, p.163].

5.1   Rutherford and Bohr

The association of Ernest Rutherford with Neils Bohr was crucial to the development of the QM paradigm. Rutherford was the premier *experimentalist* of his time; in 1908 he used alpha particles to bombard gold foil and it produced one of the most important experimental results in history. Rutherford discovered the internal structure of the atom as a densely packed, positively charged nucleus in the center with negatively charged electrons "in orbits" around it. Years later he would collaborate with a *theoretician*, Niels Bohr, to formulate the first quantized model of electron orbits. This model explained the empirical data from hydrogen spectra and it was the first real break with classical atomic mechanics [22, p.56]. But the crucial collaboration of Bohr and Rutherford was a fortuitous result of unscientific and idiosyncratic human behavior.

---

[12]Einstein would win his Nobel Prize for his photoelectric effect experiments; not for any of his theoretical work on Relativity

5.1.1 Unsatisfactory Experiences at the Cavendish Lab

In 1895, Rutherford travelled from New Zealand to work as a post-graduate at the Cavendish Lab in Cambridge. Because he was a colonial from New Zealand, his British "peers" tended to look down upon him. In 1907, he became the head of the physics department at Victoria University in Manchester, where he would perform his crucial alpha particle experiment. [22, p.39]

Then in early 1911, the theoretician Niels Bohr arrived at the Cavendish lab from the Continent. Bohr was politely greeted by the British establishment, but professionally ignored. Bohr spent months trying to engage with the head of the Cavendish lab, J. J. Thompson. But Thompson had no time for Bohr. So in late 1911, Bohr left Cambridge for Manchester.

5.1.2 Encounter at Manchester

Rutherford was a gruff experimentalist and he was usually disdainful of theoreticians. But when Bohr came to Manchester, Rutherford saw him as a comrade. Like Rutherford, Bohr had unsatisfactory experiences with the Cambridge British establishment. So a friendship was born and Rutherford became a mentor to Bohr. This relationship would lead to their collaboration on atomic structure and the Bohr model of the atom.

5.1.3 Impact of the Bohr Model

The Bohr model of the atom worked well for simple hydrogen atoms, but it failed to explain anything more complex. Two years later Bohr became the head of the physics department in Copenhagen, where he attempted to refine and extend his quantized model of atomic structure. In the end, the first working formulation of QM was derived by one of Bohr's assistants: a student named Werner Heisenberg.

5.2 Werner Heisenberg

In 1925, Werner Heisenberg completed his matrix mechanics derivation; it was arguably the first complete working model of QM. While it was subsequently superseded by wave mechanics, Heisenberg's matrix mechanics was itself the trigger for the formulation of Shrödinger's wave mechanics. The events that led to Heisenberg's involvement in physics in general, and his formulation of QM in particular, are excellent examples of how human idiosyncrasies play an key role in the emergence of science.

5.2.1 Human Idiosyncrasies Force Heisenberg to Study Physics Instead of Math

In 1919, then eighteen year old Heisenberg entered the University of Munich with the intention of studying mathematics. For this purpose, he felt it would be necessary to participate in the math seminar run by professor Ferdinand von Lindemann. Heisenberg convinced his father, who had influence in the academic community, to arrange a meeting where young Heisenberg could discuss academic plans with von Lindemann[4].

A Disastrous Meeting
The meeting was conducted in Von Lindemann's University office. Heisenberg mentioned that he was familiar with Einstein's theory of relativity and its implications for space, time and matter; to which Von Lindemann replied, "In that case you are lost to mathematics." [16] Furthermore, Von Lindemann had his pet dog sitting on his desk and it barked during the meeting. In addition to all of these obstacles, Lindemann was also hearing impaired. All of these idiosyncrasies were working against Heisenberg and he was denied admission to Lindemann's mathematics seminar [11].

Studying Physics as a Fallback
This setback forced Heisenberg to reconsider becoming a pure mathematician. At his father's suggestion, he met with Arnold Sommerfeld

and was subsequently accepted into Sommerfeld's physics seminar. So instead of pursuing pure math, as he had intended, human idiosyncrasies forced Heisenberg to study physics with one of the early pioneers of QM. It was through Sommerfeld's tutelage that Heisenberg became an excellent mathematical *physicist* [17, p.20].

6 Science in a Far From Equilibrium State

6.1 Sub-Atomic Physics in a State of Crisis

While Heisenberg pursued his studies, the aggregation of scientific inquiry around the UV catastrophe had continued. This was producing a great deal of conflicting data that engendered several competing schools of thought, or *paradigms*. So at that point in history, the science of physics was in chaos and it had reached a bifurcation point. The following quotes illuminate this period of instability:

> "[physics is passing through] contortions in its attempts to explain the simultaneous appearance of quantum and classical phenomena." John Van Vleck [8, p.159]

> "Physics is decidedly confused at the moment; in any event, it is much too difficult for me and I wish I had never heard of it." Wolfgang Pauli [8, p.162]

> "[instead of being a physicist I would] rather be a shoemaker or even an employee in a gambling casino" Albert Einstein[8, p.163].

6.2 Heisenberg and Bohr

After Heisenberg completed his doctorate, he went on to pursue post-doctoral studies in Göttingen under the supervision of Max Born. In 1922 Heisenberg attended a guest lecture on the applications of celestial mechanics to problems in QM[13]. The lecture was delivered by

---

[13]It is interesting to note that Isaac Newton had followed a similar path. Newton studied celestial mechanics in his quest to make a Philosopher's Stone.

Neils Bohr and it intrigued Heisenberg; afterwards he asked Bohr for further clarifications. Bohr was impressed by the questions, so he invited Heisenberg to study in Copenhagen for a few weeks in the spring of 1923. Bohr would later arrange for Heisenberg to work with him in Copenhagen from September, 1924 to April, 1925. This time with Bohr would later prove to be a crucial and necessary phase in the development of QM.

Rutherford's experimentalist influence on Bohr was now being passed on to Heisenberg. This was a pivotal event in the formulation of QM. Physics was divided into two main groups: empiricists and theorists. Empiricists were relying exclusively on experimental evidence and theorists were preoccupied with mathematical manipulations; but neither side was concerned with the contradictions its approach generated with regard to the other side. Heisenberg's training had been mostly theoretical, his time with Bohr's empirical approach would complete his training and enable him to synthesize the two approaches.

After finishing his training with Bohr, Heisenberg returned to assume a faculty position at Göttingen, where he was valued as the resident expert on the Copenhagen view. Soon after Heisenberg's return to Göttingen, the results of the Bothe/Geiger radiation experiment cast serious doubts upon the fundamental assumptions of the Bohr-Kramers-Slater theory of radiation. This was a major setback for the Copenhagen school and an atmosphere of depression, gloom and pessimism descended upon Bohr and his associates at Copenhagen. But isolated from the gloom in Copenhagen, Heisenberg happily continued his work in Göttingen [17, p.252].

6.3 Individual Health Idiosyncrasies and The Emergence of a New Scientific Paradigm: The Formulation of Matrix Mechanics

In June, 1925 Heisenberg's work in Göttingen was interrupted by a severe attack of hay fever. He took a two week leave of absence and traveled to the secluded rock island of Helgoland on the coast to effect

25

a cure. This change of set and setting gave Heisenberg a chance to reflect on some of Bohr's life philosophy. In particular, he recalled Bohr's comment on how 'part of the infinity seems to lie in the grasp of those who look across the sea' [17, p.249]. He took Bohr's philosophy to heart and spent much time communing with the sea.

Heisenberg's health improved and in the seclusion of Helgoland, he was able to take a fresh look at the problems facing QM. He decided to scrap all preconceived notions on quantum behavior and apply Mach's principle, which Einstein had used during the formulation of the theory of relativity [17, p.275]. According to Born, the Mach principle states:

> "...concepts and representations that do not correspond to physically observed facts are not to be used in theoretical descriptions..." [17, p..273].

On first inspection, this principle may appear to be a tautology, but the classical notions of space and time were concepts and representations that did not correspond to the physical observations of relativistic and quantum phenomena. Utilizing Mach's principle and isolated from distractions, Heisenberg worked feverishly. He made remarkable progress and in a few days, after unknowingly reinventing matrix mathematics, formulated the first rough version of what would come to be known as matrix mechanics. With his new theory in hand, Heisenberg returned to Göttingen and refined matrix mechanics with Born and Pascal Jordan. During this refinement process, they jointly discovered the Uncertainty Principle.

6.4 A Chaotic Response to the New Paradigm

The introduction of this new paradigm followed the classic pattern that Kuhn discovered; initially the new paradigm was both simultaneously accepted and rejected by the physics community. This chaotic response also follows the pattern of non-linear systems evolution in the far from equilibrium regime.

Einstein was so vehemently opposed to QM, that he encouraged Erwin Schrödinger to attempt a different approach that used Louis de Broglie's matter waves. The result was the Schrödinger Wave Equations and these worked just as well as matrix mechanics to account for the data, while also seeming to uphold continuous space and time. But subsequent analysis showed that wave and matrix mechanics were mathematically equivalent to each other: matrix mechanics treated sub-atomic phenomena as particles, while the Wave Equations treated them as waves.

7 1927 Solvay Conference: The Official Emergence of a New Scientific Paradigm

This was an epochal moment in the history of physics. The Born-Heisenberg-Bohr interpretation of quantum mechanics, subsequently referred to as the Copenhagen interpretation, was scrutinized and debated by the world's leading physicists. Because its adoption would require the abandonment of such Newtonian rationalisms as causality and space/time models, the "old guard" physicists objected to it. Einstein was their spokesman when he said:

> "An inner voice tells me that this (Copenhagen Interpretation) is not the true Jacob. The theory accomplishes a lot, but it does not bring us any closer to the secrets of the Old One. In any case, I am convinced that He does not play dice."

As the spokesman for the Copenhagen Interpretation, Bohr refuted all of Einstein's attacks [1] and ultimately prevailed; the Solvay Conference developed into an overwhelming victory for the Copenhagen Interpretation [8, p.167]. The revolution was successful and the new quantum paradigm officially emerged.

27

Part IV
Conclusion

The formulation of QM was the result of a chain of human idiosyncratic behavior. Because of the stochastic nature of the world, even if we were given prior knowledge of Von Lindemann's noisy dog and British snobbery, it would not be possible to interpolate the Heisenberg Uncertainty Principle. So the science that emerges cannot be reduced down to the idiosyncrasies of the humans that produced it. This irreducibility reflects the strongly emergent nature of scientific knowledge. The mathematical equivalence of Matrix Mechanics and Wave Mechanics is a manifestation of robustness and autonomy in emergence. Paradoxically, these robust and autonomous properties emerged from the chaos of a complex, non-linear, social system at bifurcation. So we are left with many conundrums:

1. Rutherford and Bohr (crucial collaboration on atomic structure)

   (a) If both of them were made welcome at the Cavendish Lab, would Rutherford have taken any interest in Bohr, a mere theoretician?

   (b) If Bohr had not collaborated with Rutherford, would he have been able to theorize his atomic model?

2. Heisenberg (first workable quantum theory)

   (a) If Lindemann had accepted Heisenberg into his math seminar, or if Sommerfeld had rejected him from his physics seminar, would Heisenberg ever have gotten involved in the search for QM?

   (b) If Heisenberg had never met Bohr, would he have been exposed to the empiricists view?

(c) If Heisenberg had not gone to the seclusion of Helgoland because of his hay fever, would the distractions of Göttingen have prevented him from starting over and formulating his matrix mechanics?

3. (Bifurcation I) If Einstein had not objected to the Copenhagen Interpretation and encouraged Schrödinger to formulate his wave equations, would matrix mechanics be the dominant paradigm today?

4. (Bifurcation II) If Bohr, Heisenberg or Schrödinger, had not played their respective roles: would different sub-atomic paradigms have emerged to explain QM? Or does the principle of autonomy imply that the same paradigms would have emerged, regardless of the individuals involved?

5. The quantum revolution in the 20th Century produced two different paradigms to account for the data: matrix mechanics and wave mechanics. That implies that a different "classical physics" paradigm (eg: a Leibnizian mechanics, instead of Newtonian mechanics) could have emerged in the 17th Century. But could it have significantly differed from Newtonian mechanics?

Image Credits

Fig.1: `https://opentextbc.ca/chemistryatomfirst2eopenstax/chapter/the-kinetic-molecular-theory/`
Fig.5: `https://www.cfdsupport.com/OpenFOAM-Training-by-CFD-Support/node334.html`
Fig.6: `https://www.cfdsupport.com/Ope1280px-Black_body`

References

[1] A. Aspect. Closing the door on einstein and Bohr's quantum debate. *Physics*, 8:123, 2015.

[2] M. A. Bedau. Weak emergence. *Philosophical Perspectives*, 11:375–399, 1997.

[3] D. J. Chalmers. Strong and weak emergence. *The re-emergence of emergence*, pages 244–256, 2006.

[4] M. Eckert. Werner heisenberg: controversial scientist. *Physics World*, 14(12):35, 2001.

[5] P. Flowers, K. Theopold, R. Langley, and W. Robinson. Chemistry 2e. *Houston, Texas: OpenStax*, pages 1044–1059, 2019.

[6] A. E. Gelfand and C. C. Walker. *Ensemble modeling*. Marcel Dekker, 1984.

[7] M. Haghnevis and R. G. Askin. A modeling framework for engineered complex adaptive systems. *IEEE Systems Journal*, 6(3):520–530, 2012.

[8] D. J. Kevles and L. Pyenson. The physicists: The history of a scientific community in modern america. *PhT*, 31(3):63, 1978.

[9] C. Kittel and H. Kroemer. Thermal physics. *WIT Freman: San Fransisco*, 1980.

[10] T. S. Kuhn. The structure of scientific revolutions, 1970.

[11] T. S. Kuhn and J. Heilbron. Interview of Werner Heisenberg. In *Oral History Interviews*. Niels Bohr Library & Archives, American Institute of Physics, 1962.

[12] R. B. Laughlin. *A different universe: Reinventing physics from the bottom down*. Basic Books (AZ), 2005.

[13] D. Lide. CRC handbook of chemistry and physics, 2004.

[14] B. Lindgren. *Statistical Theory*. MacMillan Publishing Co., 3 edition, 1976.

[15] P. Manneville. Dissipative structures and weak turbulence. In *Chaos — The Interplay Between Stochastic and Deterministic Behaviour*, pages 257–272. Springer, 1995.

[16] B. R. Masters. Werner Heisenberg's path to matrix mechanics. *Opt. Photon. News*, 25(7):42–49, Jul 2014.

[17] J. Mehra and H. Rechenberg. *The historical development of quantum theory*. Springer, 1982.

[18] V. Petrov, V. Gaspar, J. Masere, and K. Showalter. Controlling chaos in the Belousov–Zhabotinsky reaction. *Nature*, 361(6409):240–243, 1993.

[19] I. Prigogine. Order out of chaos. *Man's new dialogue with nature*, 1984.

[20] O. Reynolds. An experimental investigation of the circumstances which determine whether the motion of water shall be direct or sinuous, and of the law of resistance in parallel channels. *Proceedings of the Royal Society of London*, 35(224-226):84–99, 1883.

[21] M. Ruth. Evolutionary economics at the crossroads of biology and physics. *Journal of Social and Evolutionary Systems*, 19(2):125–144, 1996.

[22] C. Snow. *The Physicists*. MacMillan Publishing Co., 1981.

[23] J. C. Sprott. Complex behavior of simple systems. In *Unifying Themes in Complex Systems*, pages 3–11. Springer, 2006.

[24] J. M. Steele. Eclipse prediction in Mesopotamia. *Archive for history of exact sciences*, 54(5):421–454, 2000.

SIZING UP THE BATTERIES: MODELLING OF ENERGY-HARVESTING SENSOR NODES IN A DELAY TOLERANT NETWORK

JEREMIAH D. DENG

Department of Information Science, University of Otago, New Zealand

jeremiah.deng@otago.ac.nz

Abstract

For energy-harvesting sensor nodes, rechargeable batteries play a critical role in sensing and transmissions. By coupling two simple Markovian queue models in a delay-tolerant networking setting, we consider the problem of battery sizing for these sensor nodes to operate effectively: given the intended energy depletion and overflow probabilities, how to decide the minimal battery capacity that is required to ensure opportunistic data exchange despite the inherent intermittency of renewable energy generation.

1 Introduction

Recently, energy-harvesting wireless sensor networks (EH-WSN) [1] have become a promising technology for sensing applications. The

I would like to dedicate this article to Professor Martin K. Purvis, who introduced me to the wonderful world of queueing theory and encouraged me to brave the less-travelled roads in mobile ad hoc networks and IoT research. This little but meticulous work of mine certainly benefitted from the pleasant chats we had around philosophy, theology, programming languages, Donald Knuth, etc. I would also like to acknowledge that Dr. Sophie Zareei, whose PhD thesis Martin and I co-supervised, did some of the initial work on the same topic.

advantage of EH-WSN is obvious - batteries on the sensor nodes can be downsized due to their energy-harvesting capability, the network enjoys longer life time, eliminating the need of frequent of battery replacement, which is especially challenging for large-scale sensor deployment. However, apart from reservoirs, most renewable energy sources are intermittent in nature, which raises new challenges in designing EH-WSNs. For example, sensors may not get proper sunshine for recharging for hours, and wearable devices operated by kinetic energy will not benefit much from humans sitting for hours. This implies the necessity of using batteries to buffer the unsteady power supply from renewable energy sources.

We consider a generic EH-WSN scenario where mobile nodes are equipped with capacity-limited batteries that are powered by harvested kinetic energy; data exchange between nodes requires 1) they are within transmission range to each other; and 2) there is sufficient energy to conduct data transmission. This is in effect an EH-WSN operating as a delay-tolerant network (DTN) [15], where data transmission is opportunistic. In such a scenario, it is both important to ensure the battery size is large enough to avoid energy depletion (and hence potential failure for transmission) and energy overflow, both detrimental to the battery life.

In a previous work [17], we have examined battery sizing in terms of depletion probability and overflow probability respectively, using a coupled data and energy queue system. In this work, we intend to investigate the mathematical properties of battery size as a function regarding the operational probability requirement, and develop an algorithm to calculate the minimum battery size needed to meet the given requirements.

2   Related Work

As a performance modelling tool, queueing theory has been employed to study EH-WSNs. Gelenbe [5] first looked the modelling of an EH-sensor node using the concept of discretized energy unit called "energy packets". The arrival of these energy packets is assumed to follow a

Poisson process. A routing approach was further developed in [6]. A more general queueing model was introduced in [10], relaxing the assumption that exactly one energy packet is required to transmit a data packet. A Markovian model with data buffering was further considered in [4]. In a recent work [17] we showed that kinetic energy harvested by fitness gears discretized as energy packets can be well modelled by Poisson processes. These previous works, however, considered only static EH sensors, without involving potential intermittent connections between EH-sensor nodes due to mobility.

On the other hand, mobility has been widely investigated in ordinary wireless sensor networks and DTNs [12, 14]. Despite some counter-arguments [3], several mobility model studies [2, 8, 16, 12] suggested that two mobile nodes' encounter follows a Poisson process in mobile ad hoc networks and DTNs. There are few studies on energy harvesting networks that investigated the effects of intermittent connections [11, 13].

3   System Modelling

Notations used in this article are listed as follows:

λ_E energy packet arrival rate

λ_D data packet arrival rate

λ_C connection arrival rate

γ_D ratio λ_D/λ_C

γ_E ratio λ_E/λ_C

γ ratio λ_D/λ_E

P_{D_0} proportion of time that there is no data in the system

P_{E_k} proportion of time that system have k energy packets $k = 0, ..., K$

ρ_D utilization factor of data buffer

ρ_E utilization factor of energy buffer

α acceptable probability of energy depletion

β acceptable probability of energy overflow

K_α battery capacity decided based on α

K_β battery capacity decided based on β

$\lceil x \rceil$ ceiling, the greatest integer more than or equal to x

3.1 The queueing model

We consider a network of mobile EH-sensors. Energy harvesting leads to Poisson arrivals of energy packets (EP) with a rate of λ_E. Energy consumption occurs when there are data packets in buffer, provided that there are nodes in proximity, which is modulated by another Poisson arrival rate λ_C. Thus an Energy queue is formed at each sensor node, which can be modelled as an $M/M/1/K$, where K is the battery capacity (in terms of number of energy packets). Data packets (DP) arrive at a Poisson rate λ_D, and leave a node if there is a connection available and there is at least an energy packet in system. As memory in a sensor node is relatively cheap and less constrained, for simplicity we set no limit to the data buffer, hence allowing the data queue to be modelled by an $M/M/1$. Clearly both queues are coupled by the connection availability. Hence the Markovian packet departures in both the Energy queue and the Data queue are modulated by the connection arrival rate λ_C.

The system diagram for a sensor node is shown in Figure 1.

Similar to [10] and [6], we focus on modelling energy needed for data transmission and assume that compared with data transmission the sensing process consumes insignificant amount of energy from the battery.

It is also worth mentioning that data transmission time is much faster than energy harvesting in a node. Given the size of sensory DP and the relatively large bandwidth in a network, packet transmission

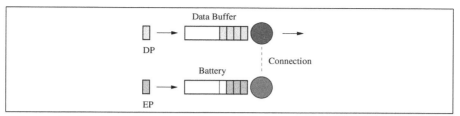

Figure 1: System diagram of a mobile sensor node. DPs arrive in the data buffer, while EPs arrive in the battery. Consumption of the energy as well as the transmission of data occur simultaneously when triggered by a connection established with another node.

time is negligible [5]. Finally, to simplify our analysis we assume at each encounter only one DP is transmitted.

These assumptions allow us to have a tractable system model with the energy and data state diagrams shown in Figure 2.

3.2 Queueing analysis

Queueing analysis has been carried out in the previous work [17]. Here we only summarize some main results.

The utilization of the data queue is given by:

$$\rho_D = \frac{\lambda_D}{\lambda_C(1 - P_{E_0})}. \tag{1}$$

For sake of system stability, we have $\rho_D < 1$. According to queueing theory, we have

$$P_{D_0} = 1 - \rho_D. \tag{2}$$

The utilization of the energy queue is

$$\rho_E = \frac{\lambda_E}{\lambda_C(1 - P_{D_0})}. \tag{3}$$

And the probability of energy depletion is

$$P_{E_0} = \frac{1 - \rho_E}{1 - \rho_E^{K+1}}, \tag{4}$$

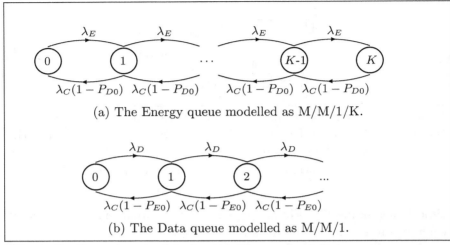

(a) The Energy queue modelled as M/M/1/K.

(b) The Data queue modelled as M/M/1.

Figure 2: Queueing models for the Energy and Data queues respectively.

while the probability of energy overflow is

$$P_{E_K} = \rho_E{}^K P_{E_0}, \tag{5}$$

By substituting Eq.(1) and Eq.(3) in Eq.(2), we have

$$P_{D_0} = 1 - \frac{\lambda_D}{\lambda_C(1 - P_{E_0})}. \tag{6}$$

Similarly, from (4), we have

$$P_{E_0} = \frac{(\lambda_C(1 - P_{D_0}))^K (\lambda_C(1 - P_{D_0}) - \lambda_E)}{(\lambda_C(1 - P_{D_0}))^{K+1} - \lambda_E{}^{K+1}}, \tag{7}$$

which, by substituting P_{D_0} using Eq.(6), becomes

$$\begin{aligned}
P_{E_0} &= \frac{\lambda_D^K(\lambda_D - \lambda_E(1 - P_{E_0}))}{\lambda_D^{K+1} - (\lambda_E(1 - P_{E_0}))^{K+1}} \\
&= \frac{\gamma^K(\gamma + P_{E_0} - 1)}{\gamma^{K+1} - (1 - P_{E_0})^{K+1}}.
\end{aligned} \tag{8}$$

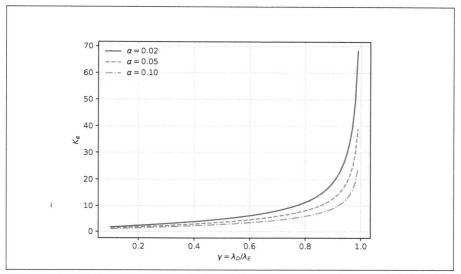

Figure 3: Different values of K_α with respect to γ for $\alpha = 0.05, 0.02$, and 0.1.

where $\gamma = \lambda_D/\lambda_E$. Here we introduce a new variable $\zeta = \frac{1-P_{E_0}}{\gamma}$, to further simplify the mathematical formulations. From the equation above, we have:

$$1 - \gamma\zeta = \frac{1 - \zeta}{1 - \zeta^{K+1}}. \tag{9}$$

Using Eq.(5), we work on the probability of energy overflow

$$P_{E_K} = \frac{(1 - P_{E_0})^K (\gamma + P_{E_0} - 1)}{\gamma^{K+1} - (1 - P_{E_0})^{K+1}}, \tag{10}$$

which can be further simplified to

$$P_{E_K} = \begin{cases} \dfrac{1 - \frac{1}{\zeta}}{1 - \frac{1}{\zeta^{K+1}}} & \zeta \neq 1, \\[2ex] \dfrac{1}{K+1} & \zeta = 1. \end{cases} \tag{11}$$

where the case of $\zeta = 1$ is obtained by using L'Hôpital's rule.

4 Battery capacity sizing

Having obtained the formulae for P_{E_0} and P_{E_K}, we are now set to find the close-form solution for the battery size as required by battery depletion and overflow probabilities. To simplify notations, let $P_{E_0} = \alpha$, $P_{E_K} = \beta$.

4.1 Battery size versus depletion probability

First we look at the battery size decided by α, denoted by K_α. K_α is in fact a function of α and γ. From Eq.(9), we have

$$1 - \gamma\zeta = \frac{1 - \zeta}{1 - \zeta^{K_\alpha+1}}, \tag{12}$$

which leads to

$$\zeta^{K_\alpha} = \frac{1 - \gamma}{1 - \gamma\zeta}. \tag{13}$$

Note that $1 - \gamma\zeta = \alpha > 0$. By taking logarithm on both sides, and substituting ζ with $\frac{1-\alpha}{\gamma}$, eventually we have

$$K_\alpha = \frac{\ln\frac{1-\gamma}{\alpha}}{\ln\frac{1-\alpha}{\gamma}}. \tag{14}$$

Note this result implies that under the required condition $\gamma < 1$, we have a positive solution of K_α. One can see that if $1 - \gamma > \alpha$, then $1 - \alpha > \gamma$; otherwise if $1 - \gamma < \alpha$, then $1 - \alpha < \gamma$. Therefore $K_\alpha > 0$. To further explore the properties of $K_\alpha > 0$, we first introduce a lemma.

Lemma 1. K_α *is monotonously increasing in terms of* γ.

The proof of Lemma 1 is given in Appendix A.1.

The interpretation is rather straightforward – the larger the γ ratio is, the more frequent DPs arrive compared with EPs, hence causing higher chance of battery depletion. To maintain the depletion probability under increased γ, a larger battery capacity is therefore needed.

From Lemma 1, we arrive at Theorem 1.

Theorem 1. *The battery size as required by the depletion probability is a monotonously decreasing function of the latter.*

Proof. Obviously, Eq.(14) contains some kind of symmetry between γ and α. Let $\alpha' = 1 - \alpha$, $\gamma' = 1 - \gamma$, then the function for calculating K_α satisfies

$$
\begin{aligned}
f(\gamma, \alpha) &= \frac{\ln(1-\gamma) - \ln \alpha}{\ln(1-\alpha) - \ln \gamma} = \frac{\ln \gamma' - \ln(1-\alpha')}{\ln \alpha' - \ln(1-\gamma')} \\
&= \frac{\ln(1-\alpha') - \ln \gamma'}{\ln(1-\gamma') - \ln \alpha'} = f(\alpha', \gamma') = f(1-\alpha, 1-\gamma).
\end{aligned}
\tag{15}
$$

This suggests that for function $f(.)$, an increased α corresponds in effect to a decreased "γ"; and an increased γ corresponds to a decreased "α". As we have already shown K_α is monotonically increasing with γ, we can now conclude K_α is a monotonically decreasing function of α. $\qquad \square$

Theorem 1 is again reasonable, since a smaller energy depletion probability would require a larger battery size. Figure 3 shows some example K_α curves with different γ and α values.

So far we have assumed $\gamma \neq 1 - \alpha$. The special case of $\gamma = 1 - \alpha$, however, is allowed, as again using L'Hôpital's rule we have

$$
K_\alpha = \frac{1}{\alpha} - 1,
\tag{16}
$$

which is also a positive, monotonically decreasing function of α.

4.2 Battery size versus overflow probability

Let K_β be the battery size decided by a given P_{E_K} value (β). We consider the normal case given in Eq.(11). Let $z = 1/\zeta$. We have

$$
\beta = \frac{1-z}{1-z^{K_\beta+1}},
\tag{17}
$$

which leads to

$$
z^{K+1} = \frac{z + \beta - 1}{\beta}.
\tag{18}
$$

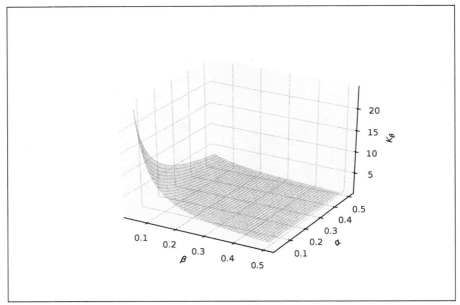

Figure 4: K_β values with respect to α and β when $\gamma = 0.95$.

Here we can see unless $z + \beta > 1$, there is no real solution for K_β. In fact, it is easy to see that when $z < 1$,

$$1 - z < \frac{1 - z}{1 - z^{K_\beta + 1}} < 1, \tag{19}$$

i.e., β has a lower bound $1 - z$.

When the condition $z + \beta > 1$ is satisfied (and naturally so when $z > 1$), we have

$$K_\beta = \frac{\ln \frac{z + \beta - 1}{\beta}}{\ln z} - 1 = \frac{\ln[\gamma - (1 - \alpha)(1 - \beta)] - \ln \beta\gamma}{\ln \gamma - \ln(1 - \alpha)} \tag{20}$$

For the battery size decided by the overflow probability, we have the following theorem:

Theorem 2. *K_β is monotonically decreasing when β increases.*

The proof of Theorem 2 is given in Appendix A.2.

Figure 4 shows the trend K_β values display across a range of α and β values when $\gamma = 0.95$.

It can be proven that for the special case of $z = 1$, i.e., $\gamma = 1 - \alpha$, we have

$$K_\beta = \lim_{z \to 1} \frac{\ln \frac{z+\beta-1}{\beta}}{\ln z} - 1 = \frac{1}{\beta} - 1. \tag{21}$$

Clearly, Theorem 2 still holds.

4.3 Battery sizing algorithm

Given different requirements in terms of α and β values, and the system setup in terms of γ, we can derive the relevant K_α and K_β to size up the battery. As seen from Eq.(20), the condition

$$(1 - \alpha)(1 - \beta) < \gamma$$

has to stand for calculating K_β. One question remains – between K_α and K_β which one actually decides the size of the battery? It is easy to see that when $\beta + \gamma = 1$, $K_\alpha = K_\beta$. Since we have shown that K_α is an monotonically increasing function of γ, and K_β a monotonically decreasing function of β, we give the following corollary without needing a formal proof:

Corollary 1. *If $\beta + \gamma < 1$, $K_\alpha < K_\beta$; if $\beta + \gamma > 1$, $K_\alpha > K_\beta$.*

Hence we have the following algorithm for battery sizing.

Algorithm 1: Battery sizing under constraints of P_{E_0} and P_{E_k}

   Data: $0 < \alpha, \beta < 1$, $(1 - \alpha)(1 - \beta) < \gamma < 1$
   Result: minimum K satisfying $P_{E_0} < \alpha$, $P_{E_K} < \beta$
1 **if** $\beta + \gamma < 1$ **then**
2 | $K \leftarrow K_\beta$ using Eq.(20);
3 **else**
4 | $K \leftarrow K_\alpha$ using Eq.(14);
5 **end**

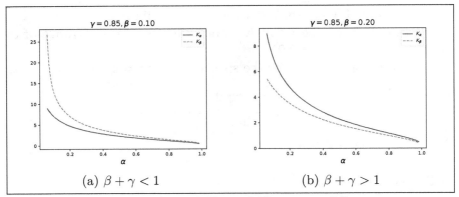

Figure 5: Comparison of K_α and K_β in two different cases.

Figure 5 gives two settings when K_α and K_β are compared. These numerical examples clearly confirm the correctness of Corollary 1.

4.4 A numerical example

To demonstrate the feasibility of an EH node in mobile settings, let us assume a scenario with the battery depletion probability $\alpha = 0.05$, and the overcharge probability $\beta = 0.3$. Suppose the three rates in the system are $\lambda_D = 0.72$ data packets/sec, $\lambda_E = 0.8$ energy packets/sec, and $\lambda_C = 0.9$ connection/sec. We have $\gamma = \lambda_D/\lambda_E = 0.9$, and $\beta + \gamma > 1$. Hence according to Algorithm 1, we obtain the required battery size $K = K_\alpha = 12.82 \approx 13$ using Eq.(14). If we use an energy packet size of $155\mu W$ which can be generated by a moderate walking activity [7], the needed battery size will be $13 \times 155\mu W = 2.015mW$.

5 Conclusion

Despite the great potential in utilizing rechargeable nodes in wireless sensor networks and body area networks, the wide application of energy-harvesting IoT systems remains elusive, largely due to the uncertain nature of energy harvesting and the lack of performance analysis results for guiding system design. In this article, we have studied the modelling of energy charging and consumption behaviours of sensor nodes in a DTN setting, where data transmission is subject to both the availability of sufficient energy, and the existence of sensor nodes in reachable proximity. A stochastic model with coupled Poisson arrival processes on the energy and data queues of the node is formulated and solved, based on which a closed-form solution of the optimal battery size is derived to meet the specified probabilities of energy depletion and overflow.

For future work, our model can be extended by considering general probability distributions for energy or data arrivals, allowing more flexible settings of data and energy packet sizes. This may enable more types of energy-harvesting sources being considered, for instance solar. Beyond using discretized energy units, continuous fluid models [9] could be employed for future investigation, where the theoretical findings may be further validated by simulation studies under realistic settings.

A Appendix – Proofs

A.1 Lemma 1

K_α is monotonically increasing on γ.

Proof. It can be worked out that

$$\frac{\partial K_\alpha}{\partial \gamma} = \frac{\frac{1}{\gamma} \ln \frac{1-\gamma}{\alpha} - \frac{1}{1-\gamma} \ln \frac{1-\alpha}{\gamma}}{\ln^2 \frac{1-\alpha}{\gamma}}.$$

We want to show that this derivative is non-negative, i.e.

$$\frac{1}{\gamma} \ln \frac{1-\gamma}{\alpha} \geq \frac{1}{1-\gamma} \ln \frac{1-\alpha}{\gamma},$$

which is equivalent to

$$\left(\frac{1-\gamma}{\alpha}\right)^{\frac{1}{\gamma}} \geq \left(\frac{1-\alpha}{\gamma}\right)^{\frac{1}{1-\gamma}}.$$

Further transformation leads to

$$(1-\gamma)^{\frac{1}{\gamma}}\gamma^{\frac{1}{1-\gamma}} \geq \alpha^{\frac{1}{\gamma}}(1-\alpha)^{\frac{1}{1-\gamma}}.$$

Let $x = 1 - \alpha$, and

$$f(z) = (1-z)^{\frac{1}{\gamma}}z^{\frac{1}{1-\gamma}},$$

so to prove the inequality above, we only need to show $f(\gamma) \geq f(x)$. To find the maximum of $f(z)$, let $f'(z) = 0$. Solving this, we get $z = \gamma$. $\qquad\square$

The derivation above also shows that the equality stands when $\gamma = 1 - \alpha$.

A.2 Theorem 2

K_β is monotonically decreasing on β.

Proof. We consider the general case where K_β can be put as

$$K_\beta = \frac{\ln(\frac{z-1}{\beta}+1)}{\ln z} - 1,$$

where $z = \frac{\gamma}{1-\alpha}$. Consider two cases only (the special case of $z = 1$ is already handled in main text):

1. $z > 1$. Both the numerator and the denominator of the fraction term are positive. Clearly the bigger β is, the smaller K_β;

2. $z < 1$. Both the numerator and the denominator are negative. With β increasing, $\frac{z-1}{\beta}$ will increase, albeit being negative. The numerator will increase as an negative value, hence the value for K_β will decrease as a positive value.

$\qquad\square$

References

[1] Kofi Sarpong Adu-Manu, Nadir Adam, Cristiano Tapparello, Hoda Ayatollahi, and Wendi Heinzelman. Energy-harvesting wireless sensor networks (eh-wsns): A review. *ACM Trans. Sen. Netw.*, 14(2), April 2018.

[2] David Aldous and Jim Fill. Reversible markov chains and random walks on graphs, 2002.

[3] A. Chaintreau, P. Hui, J. Crowcroft, C. Diot, R. Gass, and J. Scott. Impact of human mobility on opportunistic forwarding algorithms. *IEEE Transactions on Mobile Computing*, 6(6):606–620, 2007.

[4] E. De Cuypere, K. De Turck, and D. Fiems. A queueing model of an energy harvesting sensor node with data buffering. *Telecommun Systems*, 67:281–295, 2018.

[5] Erol Gelenbe. Synchronising energy harvesting and data packets in a wireless sensor. *Energies*, 8(1):356–369, 2015.

[6] Erol Gelenbe and Andrea Marin. Interconnected wireless sensors with energy harvesting. In *International Conference on Analytical and Stochastic Modeling Techniques and Applications*, pages 87–99. Springer, 2015.

[7] Maria Gorlatova, John Sarik, Mina Cong, Ioannis Kymissis, and Gil Zussman. Movers and shakers: Kinetic energy harvesting for the internet of things. *IEEE Journal on Selected Areas in Communications*, 33:1624 – 1639, 06 2015.

[8] Robin Groenevelt, Philippe Nain, and Ger Koole. The message delay in mobile ad hoc networks. *Performance Evaluation*, 62(1-4):210–228, October 2005.

[9] G. L. Jones, P. G. Harrison, U. Harder, and T. Field. Fluid queue models of battery life. In *2011 IEEE 19th Annual International Symposium on Modelling, Analysis, and Simulation of Computer and Telecommunication Systems*, pages 278–285, 2011.

[10] Yasin Murat Kadioglu and Erol Gelenbe. Packet transmission with k energy packets in an energy harvesting sensor. In *Proceedings of the 2Nd International Workshop on Energy-Aware Simulation*, ENERGYSIM '16, pages 1:1–1:6, New York, NY, USA, 2016. ACM.

[11] Mahzad Kaviani, Branislav Kusy, Raja Jurdak, Neil Bergmann, and Vicky Liu. Energy-aware forwarding strategies for delay tolerant network routing protocols. *Journal of Sensor and Actuator Networks*, 5(4):18, 2016.

[12] Amir Krifa, Chadi Barakat, and Thrasyvoulos Spyropoulos. Optimal buffer management policies for delay tolerant networks. In *2008 5th Annual IEEE Communications Society Conference on Sensor, Mesh and Ad Hoc Communications and Networks*, pages 260–268. IEEE, 2008.

[13] Yue Lu, Wei Wang, Lin Chen, Zhaoyang Zhang, and Aiping Huang. Opportunistic forwarding in energy harvesting mobile delay tolerant networks. In *Communications (ICC), 2014 IEEE International Conference on*, pages 526–531. IEEE, 2014.

[14] Paresh C Patel and Nikhil N Gondaliya. Encounter based routing and buffer management scheme in delay tolerant networks. *Encounter*, 131(18), 2015.

[15] Joel J. P. C. Rodrigues. *Advances in Delay-Tolerant Networks (DTNs) : Architecture and Enhanced Performance*. Elsevier, 2014.

[16] Thrasyvoulos Spyropoulos, Konstantinos Psounis, and Cauligi S Raghavendra. Performance analysis of mobility-assisted routing. In *Proceedings of the 7th ACM international symposium on Mobile ad hoc networking and computing*, pages 49–60. ACM, 2006.

[17] S. Zareei, A. Sedigh, J. D. Deng, and M. Purvis. Buffer management using integrated queueing models for mobile energy harvesting sensors. In *PIMRC'17*, 2017.

Resolving the 'Agency Paradox' in Agent-Based Modelling

Christopher K. Frantz

Norwegian University of Science and Technology (NTNU)

`christopher.frantz@ntnu.no`

Abstract

Agent-based modelling is a well-established modelling technique for complex social systems. With its flexibility, and increasing adoption for the modelling across diverse social-scientific disciplines, it is important to reflect on its merits and challenges, specifically with respect to the agency conception as applied in this modelling technique.

To provide a backdrop for the discussion, we first review the conception of agency entertained in agent-based modelling based on its original in multi-agent systems, contrasted with sociological debates on agency and structure. This provides a basis for highlighting the paradoxically overly contrived representation of agency in many agent-based models, alongside limited correspondence between technological and sociological perspective.

To motivate a clearer positioning of agent-based modelling for the analysis of complex social systems, we argue for the systematic inclusion of representations of institutional structure in agent-based models. Doing so, we can provide more accurate and realistic representations of existing social systems, alongside richer analytical opportunities, as well as to harmonize agent-based modelling as a technique with its underlying sociological foundations. Potential representations for institutional structure, alongside illustrative operationalisations in agent-based models are showcased to contextualise the proposal.

This text concludes with a detailed discussion of challenges and opportunities.

1 Introduction

Agent-based modelling [30] has experienced an increasingly broad uptake, leading to contributions in a diverse areas, including economics [19], physics, political science [8], and of course in the modelling of specific aspects such as religion [51]. This broader adoption not only showcases the applicability of the approach for the investigation of different facets of social phenomena and addressing disciplinary problems, but also offers grounds to reflect on its merits, as well as opportunities that agent-based modelling needs to respond to.

One of the challenges is that agent-based models, unlike other approaches, such as game-theoretical models, do not have a uniform conception of agency. This makes the development of an integrated understanding of agent-based modelling challenging, which is partly evidence of the flexibility of the concept, but, at the same time, challenges the discussion and comparison of developed models, as well as cross-disciplinary communication. Attempts for moderation, such as the use of description protocols (e.g., ODD [32, 45], ODD+D [40] and other variants) have been made and find increasing adoption, but primarily aim at documentation for the purpose of reproducability, and thus emphasise comprehensiveness. Notwithstanding those efforts, and observing the common application of the technique, we still find a great diversity in models that all claim to capture agency in widely varying form. With this observation in mind, we can reflect on 'first principles' modellers should bear in mind while engaging in the modelling process, both involving the concept of agency as used in agent-based modelling, but moreover the conceptual alignment with underlying sociological principles. Finally, we draw on principles of institutional modelling to seek for pathways that allow for more systematic development, but also to offer a more accurate representation of complex social system models.

In Section 2, following a brief introduction of principles of agent-based modelling, we will first recall the fundamental principles the agent concept is rooted in from a technological and sociological standpoint, before calling out a potentially paradox viewpoint on the representation of agency in agent-based models in Section 3. In Section 4, we then turn to proposing an alternative perspective to moderate the identified paradox by drawing on institutional concepts. In Section 5, we critically review the observations and proposed modelling approach with respect to shortcomings and opportunities, as well as providing a general concluding discussion.

2 Foundations of Agency in Agent-based Models and Sociology

2.1 Technological Perspective

From a bird's eye perspective, agent-based modelling has been around for well over half a century, with recognisable roots reaching back to Schelling's Segregation Model [47] (with pen and paper back then) that showcased the potential of simulation for social system analysis more generally, and subsequent adaptation based on use cases such as the use of micro-simulation for the purpose of tax income prediction (see e.g., [29]). The idea underlying individual-based modelling, or agent-based modelling, is that the social system is reconstructed in terms of its constituent entities. By representing behaviour as observed empirically (or postulated theoretically) in an instantiated simulation of an 'artificial society', we can generate system-level behaviour that would not have been observable on the micro level alone (i.e., on the level of individual agents).[1] Given these broad principles, agent-based modelling can be applied to various application areas such as understanding fundamental social mechanisms (e.g., cooperation [1]), offering explanation for behaviours or social mechanisms (e.g., opinion dynamics [35]), or even predicting behaviour of real societies (e.g., emergency behaviour [44]).[2]

With the plethora of application areas of agent-based modelling we can find nowadays, the role of the agent has likewise become increas-

---

[1]The concept of generative social science [18] is fundamentally associated with 'growing' social phenomena 'from the bottom up'.

[2]A broad overview on different modelling purposes is offered by [14].

51

ingly diverse. However, roots of any agency conception in the field of computational social science – the field agent-based modelling is generally associated with – necessarily point to developments in the area of multi-agent systems that have converged on the identification of proactive, reactive, autonomous and social behaviour as cornerstones of agency [62].

Of these properties, *proactivity* is characterised as the ability to deliberate about or perform actions in anticipation of action, events or otherwise changing environmental or social circumstance. *Reactiveness*, in contrast, reflects the agent's ability to respond appropriately to perceived events or invocations; the agent assumes a passive role, in contrast to the active role expected for agents with proactive behaviour, which presumes advanced capabilities, such as the ability to learn and possibly predict. *Autonomy*, as a feature associated with agency, represents the ability to act independently from its environment (e.g., without the need for external invocation).[3] A final aspect, *sociality*, is at the core of agency; drawing on human societies for inspiration, agents need to be able to mirror communicative and interactive abilities commonplace in society. All these properties themselves exist on continua, and may be of comparatively simple or complex nature: Sociality, for example, may be simplistically represented in terms of environment modifications performed by agents and sensed by others, or complex by reflecting direct communication using symbolic or higher-level linguistic structures (see e.g., [20]).

While the identification and adoption in the field of multi-agent systems has been a result of extensive debate[4], the depth of agency commonly found in agent-based models, and social simulation more generally, is typically less extensive, and by focusing on individual-level heterogeneity, multiplicity and interaction of agents [16] consequently largely emphasises reactive and social behaviour.

To motivate the discussion in the next section, but also to promote

---

[3]Noteworthy is the differentiated discussion into *executional autonomy*, in which agents can determine and control their execution cycle independently, and *motivational autonomy* that highlights the independence of deliberation an agent engages in (see [6]).

[4]Noteworthy contributions include [21], as well as [62].

prospects for refinements in agent-based modelling, the technological side alone is necessary, but not sufficient. Rather, to respond to the subject of investigation of agent-based models in computational social science – observing emergent phenomena in social systems –, it is necessary to include a sensible contextualisation from the area of sociology, given its role as theoretical basis agent-based modelling draws upon.

2.2 Sociological Perspective

Reflecting on agency from a sociological standpoint, we naturally see ourselves drawn toward the identification and delineation of social systems comprised of individual actors, as well as their environment, which naturally involves discussing the interaction between 'Structure' and 'Agency'. The associated discourse aims at resolving the question as to how social reality is produced and whether the environment or the individuals truly drive this construction of reality.

Classical proponents of the structure perspective, such as Durkheim's *structural functionalism* [13], suggest that choices of individuals are naturally defined and enabled by the societal configuration of affordances, but in consequence also constrained by social hierarchies and preimposed structure individuals are embedded in, or quite literally born into. For the sake of illustration, one may imagine a tribal community with specific rituals, conventions, and norms that are enforced by and within that very society. In analogy, imagine the same individual embedded in modern society that builds on societal arrangements that offer broad developmental opportunities, but similarly limit choice by somewhat 'objectified' legal, economic and political constraints. Advocates of structural functionalism would argue that the society an individual is embedded in frames the world view and necessitates behavioural alignment. The individual has little choice but to act within and engage with this institutional environment, individual concerns and choice are secondary.

Contrasting this perspective, proponents of Agency, such as found in flavours of *methodological individualism* [59], specifically in its later form following the Austrian School of Economics (e.g., [58]), suggest that the construction and perception of reality is subjective and can only be constructed by 'real' entities, such as individuals; concepts such as groups or associated dynamics – the Structure – are considered 'artificial' and are thus unable to create meaning in their own right. In consequence, it can only be individuals who define reality, which of course affects the perceived social structure these individuals share. This perspective is even clearer when seen through the lens of *symbolic interactionism* [39, 3] that draws attention to the micro level of individual interactions, suggesting that the observation in its purest form of individual-level interactions offers a basis to generate and understand meaning of the society more generally. Doing so, a specific feat of symbolic interactionism (in contrast to the abovementioned perspectives) is the rejection of observations on any other level than the level of individual actions, including the rejection of treatment of aggregate data about individuals (e.g., groups).

Completing this picture, later developments of post-structuralism that sought reconciliation of the extremal dichotomous positions suggesting primacy of either Agency or Structure. As such, [9] focused on interaction effects between the sociological micro level reflecting individual agents, and the macro level that captures emergent structures, that, at the same time, show reverse causal influence towards the micro level. Given the continuity of this process, Coleman's approach moderates claims about precedence of either Agency (i.e., micro-level actions) or Structure (macro-level properties). Other theorists seeking the integrated perspective include [4] and [28]. Bourdieu presumes the shaping of individuals' behaviours based on internalisation of societal structure and formation of 'habitus' based on the societal framing and individual choice, followed by the socialisation which inadvertently affects societal structure. While Giddens argues for analogous patterns of internalisation and socialisation, the central contrast is the presumption of intentionality in action choices, where Bourdieu recognises unintentional behaviour based on the brute necessity to act.

3 The Agency Paradox

Observing the brief characterisation of agency from both a technological and sociological perspective, we are prepared to resume the discussion of challenges to the agency concept as entertained in the context of agent-based modelling broadly, and the initial point of the discussion.

Contextualised with the sociological underpinnings, the reader may be quick to move Agency (over Structure) into the spotlight of agent-based modelling, and may further suggest that agents follow the principles of methodological individualism in shaping emergent structural outcomes. While subject to discussion itself (see e.g., [36, 63]), modern interpretations of methodological individualism are largely associated with rational choice approaches as commonly applied in the area of economics. While this is an area of importance in agent-based modelling (see e.g., [56]), it is more immediately associated with reductionist approaches such as game theory that emphasises abstract representations of utility. However, whether presuming the strict perspective of independent actorship, or shifting to a post-structuralist perspective – a harmonised and integrated consideration of Agency and Structure – , an aspect that challenges agent-based modelling as a technique is the generally limited, and at best ascriptive conception of Agency in many models that foregoes the representation of cognitive abilities and focuses on properties of sociality based on interaction (which further implies the consideration of reactiveness).[5] Explanations for this contrived representation, as indicated before (see Section 2; [16]), in part lie in the focus on emergent phenomena in agent-based models, which requires simulation of large number of entities and limits the complexity of simulated individuals. Traditionally, this has been associated with constraints in computational power (an argument of limited validity in modern times) as well as the limited support by well-established simulation tools (e.g., NetLogo [60]).[6] An-

---

[5]This does, of course, not apply to all agent-based models. Exceptions, such as [5] are noteworthy, but more commonly found in theoretical contributions, rather than empirically-grounded models.

[6]It is noteworthy to clarify that there exist a plethora of different simulation

other aspect of methodological nature is the promotion of the KIDS approach [15] for empirically-grounded models that drives the focus on descriptively detailed models 'close to the domain' as well as analytical necessities. This is complementary to the broad interdisciplinary adoption of agent-based modelling as outlined in the beginning, which invites for agency of varying complexity and naturally challenges the accommodation of conceptual abstractions outside of the modeller's core domain, let alone attempting a more comprehensive representation of human-like agency.[7]

This observation leads to the identification of circumstances that one could characterise as paradox: while emphasising the role of agency in reconstructing complex social scenarios and associated phenomena of analytical interest, the tailored agency conception is oftentimes overly contrived, largely (and often merely) focusing on the specification of essential individual- or group-level characteristics, but goes little beyond basic reflections of sociality and reactive behaviour. Where existent, cognitive abilities are limited both with respect to deliberation capabilities[8] as well as social-psychological abilities, despite calls for a more comprehensive consideration (see e.g., [37]) so as to consider the representation of sociality in agents as meaningful in the first place. Conversely, where cognition is too limited, a truthful representation of second-order emergence – the interaction with emergent phenomena on the part of the agents – cannot be sensibly developed.

While this observation suggests that agent-based models at large are unable to accurately reflect the individual-level behaviour the modelled real-life entities exhibit, this shall not be misread as a call to systematically and comprehensively embed cognitive abilities. More importantly, it demands for a clear justification of agent properties in given models, and seek correspondence to the complexity of the environment the model captures, hence being "cognitively plausible" [17]

---

tools. Those vary, however, in complexity, focus on target audience, and disciplinary accessibility.

[7]Amongst the most comprehensive approaches that aim at representing human capabilities with great detail include CLARION [55] and SOAR [38].

[8]Selected approaches are highlighted by [2].

within the reference frame of the developed simulation model.[9] Operationally, this invites for a more systematic and precise specification of the nature of agency a model embeds, an aspect documentation protocols increasingly account for (e.g., [40, 33]). Furthermore, models with overly restricted agency conceptions should further guard themselves against any overgeneralisation of interesting outcomes of such models may produce.

4 Towards a Minimal Agency Conception in Agent-based Models

Beyond the challenging conception of Agency, the discussion of the sociological Agency conception, and its intimate entanglement with conceptions of Structure invites for suggestions that in part alleviates the concern highlighted above, but may further guide modellers both during model development and its documentation.

Since agent-based models are primarily concerned with "growing Structure" based on interaction of individuals that ideally reflect the observed society in quantity, a perhaps more constructive pointer is the concern with structure. While emergence of structure is of central concern in agent-based models, it is not only outcome; the simulation model itself, including its parameterisation, modelled environment, social organisation, etc. is 'Structure' and input to the model at the same time.[10] Hence, accepting that Structure is both input and outcome of agent-based models, a sensible call is to propose a systematic consideration of fundamental coordinative structures as part of any modelling effort.

But how one can model or represent concepts that (ideally) only emerge as part of the simulation outcome? A possible response can be drawn by borrowing from the real world in the form of the structure

---

[9]'Cognitively plausible' is to be understood in contrast to 'cognitively accurate'.

[10]Borrowing from the area of new institutional economics, when identifying the smallest unit of analysis and the onset of any coordinative structure in economic relationships, Williamson famously stated "at the beginning there were markets" ([61], p.20) which of course are institutional structures in their own right.

that frames our actions and coordinates our behaviours: *Institutions.*

4.1 Institutions as 'Structure'

Institutions, referred as the 'rules of the game' [41], provide the basis of any coordinative structure individuals underlie, including behavioural conventions and norms that capture the socio-institutional strata of institutions, as well as regulation, laws and other forms of formally adjudicated rules that occupy the formal-institutional position in the institutional spectrum [57]. Doing so, institutions realise the foundation for coordinated behaviour; the sociality component of Agency. It takes only little reflection to then characterise the belief in institutions as the lowest common denominator of any society – which should thus find correspondence in any model of human society. Complementing the call for stronger consideration of socio-psychological concepts in agent-based modelling [37], it comes at little surprise to argue for similar considerations from the sociological perspective, treating 'minds [themselves] as social institutions' [7].

Suggesting the systematic inclusion of institutional concepts in agent-based modelling, we need to expand on the scope of such institutional modelling. Institutions, as defined by [41] (rules of game), [31] (behavioural regularities and belief systems), and [43] (rules to govern interaction within social systems) emphasise the regulative character conventionally associated with institutions, that is, their ability to coordinate behaviour and respond to transgressions. However, inasmuch as simulation scenarios consist of a social, they may also exist of a biophysical environment that embeds the modelled entities, or rather – as with the agency conception – the aspects of such system that are of analytical relevance. Here the constitutive role of institutions is of central concern, that is the population and configuration of an institutional scenario by introducing, defining, or otherwise modifying organisational structure and entities, such as objects, artefacts, roles or actions in the model that are subsequently engaged in or otherwise referred to in the scenario.

While the regulative side has found considerable attention both in modelling and representation in institutional models, such as [54,

27, 25], to date the constitutive side has found limited consideration, despite discussion in the legal and logical domain (see e.g., [46], [50], [34]).

In its crudest form, constitutive rules can be described as X counts as Y in context Z [50], where X represents an entity of physical of virtual nature, Z a set of circumstances, and Y the role, function or interpretation X holds under such circumstances.

Illustrating this concept with reference to an imaginary organisational structure, such as found in corporations, we will the specification of roles within that organisation and their interrelationships defined in constitutive terms, such as a CEO, logistics manager or warehouse labourer. This includes the specification of hierarchy (subordination of employees under CEO), but also the general rights individuals in such roles hold (e.g., right to fair treatment in employment relationships, right to representation in trade unions, etc.). In this exemplified setting, a set of actions may have a distinctive meaning, such as 'being employed', or 'being fired', whose meaning and consequence may vary under different circumstances (e.g., different labour law regimes), or simply may not hold at all (e.g., personal life). However, beyond the actor-centric specification of structure, constitutive rules can define objects or artefacts of operational relevance, such as a 'bill of lading' or 'notification or dispatch', which may of course be subject to more general regulatory specification, but may carry specific provisions in the imagined organisational context.

In as far as constitutive rules capture roles, object, artefacts and objects, they thus 'set the scene' by populating the necessary structure in terms of actors, actions, and other entities of relevance. Regulative rules, in contrast, provide the mechanism to afford the specification of associated obligations, prohibitions or permissions in operational terms – in short: the *Do*s, *Don't*s and *May*s. Returning to our examples, they include the duties a CEO is subjected to (e.g., reporting duties), but also under which conditions a 'notification of dispatch' is to be given (however, not its form or contextual meaning – which is specified in constitutive terms).

With this compound interpretation of institutions, we have a basis to capture structure not only in regulative terms, but moreover offer a vehicle to specify 'structure' of simulation scenarios in a configurational sense. This allows us to review the incongruent consideration of Agency and Structure as set out earlier in this section, and essentially reconcile the sociological underpinnings with the scientific role agent-based models hold for the purpose of social systems analysis: Agency cannot be modelled without concern for Structure, hence a minimal model of Agency needs to consider (Institutional) Structure.

4.2   Integrating Agents and Institutions

Offering this reasoning, we are left with a final challenge. A central feat of human social systems[11] (and, in consequence, agent-based models) is the notion of 'immergence', that is, the recognition and reaction, and potential interaction with the emergent phenomenon. As individuals, for example, we are able to recognise formations of groups or other patterns of social arrangements that transcend individuals' activities and bear emergent properties. In order to interact with such entities, we require mechanisms to capture and articulate the structures we observe.[12] While generally captured in our ability to communicate the existence of such structures, this has immediate consequences for agent-based models, and the agent concept more immediately.

If we expect agents to recognise and respond to emergent structural properties, the consideration of institutional representations as presumed minimal cognitive ability becomes imperative. However, while doing so implies a richer and systematic consideration of cognitive concepts, it is important to recognise the complexity trade-off of the associated agent concept. Inasmuch as the complexity of the agent concept responds to analytical needs (let alone available empirical information), the institutions concept (and, in turn, its cognitive

---

[11]At this stage, we will not engage in the debate as to whether and to which extent the same holds for non-human social systems – while the author suggests it likely does.

[12]Noteworthy at this stage is Searle's explicit reference to language [49] as precondition for recognising and engaging in institutional arrangements.

representation) needs to mirror this complexity. We shall thus not entertain the claim to promote specific notions of institutional complexity, but rather suggest *that* they find systematic representation, and do so in congruence with the agent model.

While sensibly motivated to this stage, to lend instructive insights and stimuli for further development, as a final substantive aspect of this paper, we will turn to the discussion of and pointers to specific approaches that can serve as a basis to reconcile the conceptual embedding of Structure within Agency. To this end, we are left to turn to existing and novel approaches in the area of political science more generally, and institutional analysis more specifically, as well as computational social science, to suggest potential pathways.

Out of the many approaches applied to date,[13] we will highlight a specific approach that has found broader recognition, likely due to its interdisciplinary accessibility and empirical foundation, the 'Grammar of Institutions' by [10, 11].

The Grammar of Institutions, more recently (and in the remainder of this work) referred to as *Institutional Grammar* (IG), has been proposed in the context of institutional analysis, and more specifically, the analysis of self-governed Common-Pool Resource (CPR) regimes, in which stakeholders, absent external enforcement, govern the fair exploitation of shared resources,[14] with general reference to natural resources such as pastures, forestry, or fishing grounds.[15] Embedded in the central framework for institutional analysis, the Institutional Analysis and Design Framework (IAD) [42], the IG serves as a mechanism to express the rules enacted by the analysed communities. The essential structure of the IG includes the characterisation of actors, alongside actions and potentially affected objects regulated in the form of obligations, prohibitions or permissions within a specified context,

---

[13]Richer overviews of existing alternatives – albeit largely from the technical perspective – are discussed in [12] and [23].

[14]Fair redistribution is of course specified and governed with respect to the specific arrangement and in itself subject to analysis.

[15]While empirically based on the management of environmental resources, the analytical principles equally apply to other resources, such as commons-based peer production in the software development context [48].

i.e., under certain conditions. This is expressed in syntactic form (with symbolic labels italicised) as 'Attribute' – representing actor attributes in IG parlance –, 'oBject' – potentially affected objects[16] –, 'Deontic' – reflecting the associated discretion or regulation of the specified action –, 'aIm' – reflecting the action itself –, 'Conditions' – reflecting the characterisation of circumstances under which the regulation applies –, and the 'Or else' component that captures potential sanctions for non-compliance.

Using this basic syntactic form, any self- or externally imposed constraints on individuals' behaviour can be uniformly specified, such as 'CEO (A) must not (D) fire (I) employees (B) without prior notification (C), or else (union representative (A) may (D) lodge (I) a formal complaint (B)) (O).' Note that this example goes beyond the original syntactic coding of the IG, but further considers conceptions of nesting[17] (signalled by parentheses) that reflect the interdependency of action relationships between actors as well as higher-level complexity of institutional configurations.

While the IG captures the regulative side of institutional configurations that offers the capability to specify operational constraints in the form of such so-called institutional statements [10], the IG does not consider the constitutive side of institutional structure. To this end – and resolving the gap of providing a facility to introduce scenario-relevant structural information, a recent proposal for refinement, IG 2.0 [26], offers an explicit representation of constitutive statements. This approach essentially recognises the existence of constituted entities (E), constitutive functions (F), and constituting properties (P) amongst context characterisations and specification of consequences following the mechanisms of the original institutional grammar, thus harmonising and integrating 'both sides of institutional structure'. An exemplified representation[18], referencing the same illustrative domain, one may surmise that 'The CEO (E) is (F) the top-level position

---

[16]This syntactic component was subsequently introduced by [52].

[17]This conception, tagged Nested ADICO, has been subsequently introduced by [22].

[18]The following encoding is contrived, but showcases the essential coding of constitutive statements.

(P) within a corporate structure (C).', thus offering the definition of a role without making immediate operational reference or otherwise constraining behaviour. Constitutive statements, however, also offer further characterisations of the role in terms of endowed rights, e.g., 'The CEO (E) has (F) the right to take operational decisions related to business operations, including production, employment and logistics (P).', or definition of other relevant entities (e.g., 'A notification of termination (E) contains (F) the following information: ... (P)').

With this ability to capture both regulative, that is specifications of behavioural constraints and definitional/existential characterisations, we are in a position to suggest that observable behavioural phenomena, whether directly associated with individual agents or environmental circumstances, can be – in principle – uniformly represented without necessarily requiring a prior conception, and potentially be result of generative production.

What is left to discuss is the reconciliation of 'Agency' and 'Structure' – irrespective of their complexity – in agent-based models. Instead of erring on the side of complexity and cognitive ability by suggesting the need to afford deliberation and reasoning akin to architectures found in multi-agent systems, the need to reestablish congruence between agent-based modelling and its social-scientific (and more specifically sociological underpinnings) suggests a minimalist approach that relies on the functional embedding of institutions – responding to the scenario-dependent complexity – in agent architectures. Agents thus need to be able identify and operate on institutional representations, both to respect those as coordinative basis of social systems. Epistemologically more important from a perspective of agent-based models, introducing an explicit representation provides the basis to meaningfully respond to emergent properties, and where sensible and realistic, interact with those on the basis on representations that establish both cognitive as well as functional correspondence to the coordinative and constitutive structure found in the real world.

Approaches for agent architectures and agent-based models that consider such minimal representations exist. Notable early work in the area of agent-based modelling and the field of environmental resource management include [54, 53], who modelled a CPR water management regime and endogenously 'populating' rules for levels of exploitation based on learning and behavioural prediction, and expressed those in terms of institutional statements as described above. Another work by [25] extended the generative approach to the 'construction' of behavioural rules based on experiential learning in an abstract hypothetical trade scenario. Both approaches, however, operate on regulative statements only and do not embrace the characterisation of structure in constitutive terms. Nevertheless, existing work in this area suggests that an integrated conception of agents with institutions of corresponding complexity can be realised. Doing so, one can suggest, we can recover the unique proposition that agent-based modelling, or rather *agent-based institutional modelling*, can make over other, more expressive and sophisticated notions of agency: it is the idiomatic approach to simulate complexity found in human social systems by equally committing to the modelling of individuals, and with no less concern, the structure they find themselves embedded in.

5 Discussion

Before we turn to a concluding discussion, we briefly summarise essential aspects arguments laid out above.

5.1 Summary

In this work, we explored limited correspondence of agency conceptions with individuals they represent, discussed in form of the paradox situation that agent-based models aim at exploring emergence effects of complex social systems in detail, specifically if compared to reductionist approaches that aim at more abstract representations. However, reflecting on existing conceptions of Agency commonly found in agent-based models, we recognise that those satisfy the principles of Agency – established in the context of multi-agent systems as conceptual ancestor – only to a limited extent, potentially fueling suggestions that agent-based modelling is unable to offer a valid representation of the entities agents are meant to model.

Offering the counter position that agent-based models rely on 'interaction over cognition' to produce emergent macro-level outcomes, a separate observation offers an opportunity to review and reflect on agent-based modelling and the sociological foundations relevant for any analysis of social systems – more specifically, the systematic consideration of structure in agent-based models.

Emphasising the role institutional aspects for the governance of interaction, and thus for a minimalist conception of Agency in agent-based models, a specific emphasis is put on institutional structures, and moreover, reflecting both such structure in regulative terms, i.e., controlling behaviour, and constitutive terms, i.e., configuring an interaction scenario with respect to embedded entities.

The reintegration of Agency and structural representations are showcased using a institution conception well established in the context of political science and public policy analysis – the Institutional Grammar – , alongside further refinements that facilitate the representation of both constitutive and regulative sides of institutions.

As a final aspect, this work made brief reference to existing works in the area of agent-based modelling that offer an operational integration of institution conceptions in agent-based models and are able to produce emergent structural outcomes as part of the model execution.

5.2   Discussion

We are left to discuss the benefits of such approach. Attempting the integration of Agency and Structure, we can provide multiple benefits.

Firstly, at the current stage, in many respects agent-based models are more strongly defined by the development and simulation platforms they are implemented in, rather than the methodologies applied. And while documentation standards exist, their specificity with respect to structural aspects is limited, thus hampering comparability of models.

Secondly, agent-based models vary vastly in scale, complexity and sophistication. A specific challenge here is that models are described and designed with varying levels of depth with respect to the involved entities. Offering, if not recommending, an explicit representation for structural aspects of the model as first-order entities may promote stronger attention to such aspects in the model – and tacitly afford the necessary documentation.

A final benefit is the facilitation of a more sophisticated consideration of immergence, that is, the ability of individuals to respond to or interact with properties that emerge in simulation models in the first place. By affording an explicit representation, the extent to which agent-based models can meaningfully reflect existing human societies in terms of sophistication and nuance would greatly improve. Providing explicit representations offers a further central benefit, such as fostering explanatory potential by drawing on the simulation, or even agents themselves, to offer an explanation of emergent properties (see [24]).

A final aspect of concern is the epistemological grounding. As with every field, applied techniques need to maintain representational correspondence to the underlying theory. While possibly less structured than many other fields, computational social science fundamentally builds on sociological principles and should thus maintain a conceptual link. Only by maintaining such ties, can agent-based modelling retain applicability to fields, scenarios and theory that rely on these first principles. More importantly, however, the theoretical correspondence would resolve any discourse around the 'agency paradox' as characterised in this work by developing a clear minimal conception of agency. Seeking alignment would further provide agent-based modelling with a more explicit 'theoretical home', and, while doing so, offer a proposition to strengthen its unique ability to model complex social systems.

With these opportunities and motivations in mind, we of course need to consider challenges that arise from this discussion, specifically of methodological and operational nature. Broader methodological concerns involve the adaptation and development of guidelines and procedures that afford the systematic and integrated consideration of agent-based institutional modelling without 'overfitting' to specific social system arrangements or fields.

A noteworthy concern is related to scope of the discussion. In this work, Structure has been identified in the institutional sense, i.e., all aspects of the system that are of coordinative nature or carry meaning from a subjective perspective. Aspects of the bio-physical system, its interactions and processes, as essential for the purpose of socio-ecological systems modelling specifically, – essentially brute facts and physical laws – have been omitted from the discussion. While those are of relevance, and may affect the institutional reality as represented in social systems, their relationship to Structure in the sociological sense is an important consideration, but outside the scope of the discussion offered in this work.

Another aspect is the consideration of practical concerns, both with respect to implementation as well as execution, leading to the following questions: If we require explicit representations of structural aspects of a system, a) in how far does this challenge runtime scalability of such model, and b) in how far does this affect the development of such model?

While the critical reader may of course suggest improvements of the empirical grounding of selected claims made in this work (e.g., the common use of rather simple models of agency), these statements are based on the author's acquaintance with modelling as it is commonly entertained in the field. An empirical exploration is certainly desirable, but beyond the scope of this work. A further noteworthy clarification is certainly that institutions, organisations, and other co-ordinative features have found reflections in agent-based modelling, in some instance engaging entire subcommunities, specifically in the multi-agent systems field (e.g., electronic institutions, organisational multi-agent systems, normative multi-agent systems). The essential point made here, however, is that such principles should be applied in general modelling practice, across various application fields and beyond specific communities.

At this stage, we reach the conclusion of the discussion. While this work can be read as a proposal for a transformative change in modelling practice, it foremostly intends to raise awareness about the importance of maintaining a) conceptual correspondence between the fundamental theory and its application in techniques, and b) more specifically, the entanglement of individual and the environment it is embedded in. The consideration of both earlier points may render a more accurate and expressive representation of the modelled social system, and moreover, the emergent phenomena. Whatever the reader's call on precedence of either Agency or Structure, an undoubtedly central concern is the relationship that exists between both – the fact that *either one necessitates the other*. But if such is the case, why should *only one* of those intertwined concepts find explicit representation when modelling complex social systems?

References

[1] Robert Axelrod. An Evolutionary Approach to Norms. *The American Political Science Review*, 80(4):1095–1111, 1986.

[2] Tina Balke and Nigel Gilbert. How Do Agents Make Decisions? A Survey. *Journal of Artificial Societies and Social Simulation*, 17(4):13, 2014.

[3] H. Blumer. *Symbolic Interactionism: Perspective and Method*. Prentice Hall, 1969.

[4] Pierre Bourdieu. *An Outline of a Theory of Practice*. Cambridge University Press, London, 1977.

[5] George Butler, Gabriella Pigozzi, and Juliette Rouchier. Mixing dyadic and deliberative opinion dynamics in an agent-based model of group decision-making. *Complexity*, 2019:1–31, 08 2019.

[6] C. Castelfranchi. Guarantees for autonomy in cognitive agent architecture. In Michael J. Wooldridge and Nicholas R. Jennings, editors, *Proceedings of the Workshop on Agent Theories, Architectures and Languages, ATAL '94*, volume 890 of *Lecture Notes in Artificial Intelligence*, pages 56–70, Berlin, 1995. Springer.

[7] Cristiano Castelfranchi. Minds as social institutions. *Phenomenology and the Cognitive Sciences*, 13(1):121–143, 2014.

[8] Lars-Erik Cederman. Computational models of social forms: Advancing generative process theory. *American Journal of Sociology*, 110(4):864–893, 2005.

[9] James S. Coleman. *Foundations of Social Theory*. Harvard University Press, Cambridge (MA), 1990.

[10] Sue E.S. Crawford and Elinor Ostrom. A Grammar of Institutions. *The American Political Science Review*, 89(3):582–600, September 1995.

[11] Sue E.S. Crawford and Elinor Ostrom. A Grammar of Institutions. In *Understanding Institutional Diversity*, chapter 5, pages 137–174. Princeton University Press, Princeton (NJ), 2005.

[12] Karen da Silva Figueiredo, Viviane Torres da Silva, and Christiano de Oliveira Braga. Modeling Norms in Multi-agent Systems with NormML. In Marina De Vos, Nicoletta Fornara, JeremyV. Pitt, and George Vouros, editors, *Coordination, Organizations, Institutions, and Norms in Agent Systems VI*, volume 6541 of *Lecture Notes in Computer Science*, pages 39–57. Springer, Berlin, 2011.

[13] Émile Durkheim. *The Division of Labour in Society*. Free Press, New York (NY), 1933. Originally published: 1893.

[14] Bruce Edmonds. *Different Modelling Purposes*, pages 39–58. Springer International Publishing, Cham, 2017.

[15] Bruce Edmonds and Scott Moss. From KISS to KIDS – An 'Anti-simplistic' Modelling Approach. In Paul Davidsson, Brian Logan, and Keiki Takadama, editors, *Multi-Agent and Multi-Agent-Based Simulation*, volume 3415 of *Lecture Notes in Computer Science*, pages 130–144. Springer, Berlin, 2005.

[16] Corinna Elsenbroich and Nigel Gilbert. *Modelling Norms*. Springer, Dordrecht, 2014.

[17] Joshua Epstein. *Toward Neurocognitive Foundations for Generative Social Science*. Princeton University Press, 2014.

[18] Joshua M. Epstein. *Generative Social Science: Studies in Agent-Based Computational Modeling*. Princeton University Press, Princeton (NJ), 2007.

[19] J. D. Farmer and D. Foley. The economy needs agent-based modelling. *Nature*, 460:685–686, 2009.

[20] Jacques Ferber. *Multi-Agent Systems - An Introduction to Distributed Artificial Intelligence*. Addison-Wesley, Harlow (UK), 1999.

[21] S. Franklin and A. Graesser. Is it an Agent, or just a Program?: A Taxonomy for Autonomous Agent. In *Proceedings of the Third International Workshop on Agent Theories, Architectures, and Language (ECAI)*, pages 21–25, 1996.

[22] C. Frantz, M. K. Purvis, M. Nowostawski, and B. T. R. Savarimuthu.

nADICO: A Nested Grammar of Institutions. In G. Boella, E. Elkind, B. T. R. Savarimuthu, F. Dignum, and M. K. Purvis, editors, *PRIMA 2013: Principles and Practice of Multi-Agent Systems*, volume 8291 of *Lecture Notes in Artificial Intelligence*, pages 429–436, Berlin, 2013. Springer.

[23] C. K. Frantz and G. Pigozzi. Modeling norm dynamics in multi-agent systems. *IfCoLoG Journal of Applied Logic*, 5(2):491–564, 2018.

[24] Christopher K. Frantz. Unleashing the agents: From a descriptive to an explanatory perspective in agent-based modelling. In Harko Verhagen, Melania Borit, Giangiacomo Bravo, and Nanda Wijermans, editors, *Advances in Social Simulation*, pages 169–185, Cham, 2020. Springer International Publishing.

[25] Christopher K. Frantz, Martin K. Purvis, Bastin Tony Roy Savarimuthu, and Mariusz Nowostawski. Modelling Dynamic Normative Understanding in Agent Societies. *Scalable Computing: Practice and Experience*, 16(4):355–378, 2015.

[26] Christopher K. Frantz and Saba Siddiki. Institutional Grammar 2.0: A specification for encoding and analyzing institutional design. *Public Administration*, 99:222–247, 2021.

[27] Amineh Ghorbani, Pieter Bots, Virginia Dignum, and Gerard Dijkema. MAIA: a Framework for Developing Agent-Based Social Simulations. *Journal of Artificial Societies and Social Simulation*, 16(2):9, 2013.

[28] Anthony Giddens. *The Constitution of Society*. Polity Press, Cambridge (UK), 1984.

[29] N. Gilbert and K. G. Troitzsch. *Simulation for the Social Scientist*. Open University Press, Maidenhead, 2005.

[30] Nigel Gilbert. *Agent-Based Models*. Sage Publications, London, 2008.

[31] Avner Greif. *Institutions and the Path to the Modern Economy: Lessons from Medieval Trade*. Cambridge University Press, New York (NY), 2006.

[32] Volker Grimm, Uta Berger, Finn Bastiansen, Sigrunn Eliassen, Vincent Ginot, Jarl Giske, John Goss-Custard, Tamara Grand, Simone K. Heinz, Geir Huse, Andreas Huth, Jane U. Jepsen, Christian Jørgensen, Wolf M. Mooij, Birgit Müller, Guy Pe'er, Cyril Piou, Steven F. Railsback, Andrew M. Robbins, Martha M. Robbins, Eva Rossmanith, Nadja Rüger, Espen Strand, Sami Souissi, Richard A. Stillman, Rune Vabø, Ute Visser, and Donald L. DeAngelis. A standard protocol for describing individual-based and agent-based models. *Ecological Modelling*,

198(1–2):115–126, 2006.

[33] Volker Grimm, Steven F. Railsback, Christian E. Vincenot, Uta Berger, Cara Gallagher, Donald L. DeAngelis, Bruce Edmonds, Jiaqi Ge, Jarl Giske, Jürgen Groeneveld, Alice S.A. Johnston, Alexander Milles, Jacob Nabe-Nielsen, J. Gareth Polhill, Viktoriia Radchuk, Marie-Sophie Rohwäder, Richard A. Stillman, Jan C. Thiele, and Daniel Ayllón. The odd protocol for describing agent-based and other simulation models: A second update to improve clarity, replication, and structural realism. *Journal of Artificial Societies and Social Simulation*, 23(2):7, 2020.

[34] Davide Grossi, John-Jules Ch. Meyer, and Frank Dignum. The many faces of counts-as: A formal analysis of constitutive rules. *Journal of Applied Logic*, 6(2):192 – 217, 2008. Selected papers from the 8th International Workshop on Deontic Logic in Computer Science.

[35] R. Hegselmann and U. Krause. Opinion dynamics and bounded confidence models, analysis, and simulation. *Journal of Artificial Societies and Social Simulation*, 5(3):2, 2002.

[36] Francesco Di Iorio and Shu-Heng Chen. On the connection between agent-based simulation and methodological individualism. *Social Science Information*, 58(2):354–376, 2019.

[37] Wander Jager. Enhancing the realism of simulation (eros): On implementing and developing psychological theory in social simulation. *Journal of Artificial Societies and Social Simulation*, 20(3):14, 2017.

[38] J. E. Liard. *The Soar Cognitive Architecture*. MIT Press, 2012.

[39] G.H. Mead. *Mind, Self and Society*. University of Chicago Press, Chicago, 1934.

[40] Birgit Müller, Friedrich Bohn, Gunnar Dreßler, Jürgen Groeneveld, Christian Klassert, Romina Martin, Maja Schlüter, Jule Schulze, Hanna Weise, and Nina Schwarz. Describing human decisions in agent-based models – odd + d, an extension of the odd protocol. *Environmental Modelling & Software*, 48:37 – 48, 2013.

[41] Douglass C. North. *Institutions, Institutional Change, and Economic Performance*. Cambridge University Press, New York (NY), 1990.

[42] E. Ostrom, R. Gardner, and J. Walker. *Rules, Games, and Common-Pool Resources*. University of Michigan Press, Ann Arbor (MI), 1994.

[43] Elinor Ostrom. *Understanding Institutional Diversity*. Princeton University Press, Princeton (NJ), 2005.

[44] Xiaoshan Pan, Charles S. Han, Ken Dauber, and Kincho H. Law. A multi-agent based framework for the simulation of human and social

behaviors during emergency evacuations. *AI & Society*, 22(2):113–132, Nov 2007.

[45] G. Polhill. ODD updated. *Journal of Artificial Societies and Social Simulation*, 13(4):9, 2010.

[46] John Rawls. Two concepts of rules. *The Philosophical Review*, 64:3–32, 1955.

[47] Thomas C. Schelling. Dynamic Models of Segregation. *The Journal of Mathematical Sociology*, 1(2):143–186, 1971.

[48] Charles M. Schweik and Robert C. English. *Internet Success – A Study of Open-Source Software Commons*. MIT Press, 2012.

[49] John Searle. Constitutive rules. *Argumenta*, 4(1):51–54, 2018.

[50] John R. Searle. *Speech Acts: An Essay in the Philosophy of Language*. Cambridge University Press, London, 1969.

[51] F. LeRon Shults, Ross Gore, Wesley J. Wildman, Christopher Lynch, Justin E. Lane, and Monica Toft. Mutually escalating religious violence: A generative model. In *Proceedings of the 2017 Social Simulation Conference, Dublin, Ireland, 2017*, 2017.

[52] Saba Siddiki, Christopher M. Weible, Xavier Basurto, and John Calanni. Dissecting Policy Designs: An Application of the Institutional Grammar Tool. *The Policy Studies Journal*, 39(1):79–103, 2011.

[53] A. Smajgl, L. Izquierdo, and M. G. A. Huigen. Rules, Knowledge and Complexity: How Agents Shape their Institutional Environment. *Journal of Modelling and Simulation of Systems*, 1(2):98–107, 2010.

[54] Alex Smajgl, Luis R. Izquierdo, and Marco Huigen. Modeling endogenous rule changes in an institutional context: The ADICO Sequence. *Advances in Complex Systems*, 2(11):199–215, 2008.

[55] Ron Sun. Cognitive social simulation. In *Anatomy of the Mind: Exploring Psychological Mechanisms and Processes with the Clarion Cognitive Architecture*. Oxford Scholarship, 2016.

[56] Leigh Tesfatsion. Agent-based computational economics: A constructive approach to economic theory. In Leigh Tesfatsion and Kenneth L. Judd, editors, *Handbook of Computational Economics*, volume 2, pages 831–880. North Holland, Amsterdam, 2006.

[57] Raimo Tuomela. *The Importance of Us: A Philosophical Study of Basic Social Notions*. Stanford University Press, Stanford (CA), 1995.

[58] Friedrich A. von Hayek. *Law Legislation and Liberty*, volume 1. University of Chicago Press, Chicago (IL), 1973.

[59] Max Weber. *Economy and Society*. University of California Press, Berkeley (CA), 1978. Originally published: 1922.

[60] Uri Wilensky. Netlogo, 1999.

[61] Oliver E. Williamson. *Markets and Hierarchies, Analysis and Antitrust Implications: A Study in the Economics of Internal Organization*. Free Press, New York (NY), 1975.

[62] Michael Wooldridge and Nicholas R. Jennings. Intelligent agents: theory and practice. *The Knowledge Engineering Review*, 10(2):115–152, 1995.

[63] Julie Zahle and Harold Kincaid. Agent-based modeling with and without methodological individualism. In Harko Verhagen, Melania Borit, Giangiacomo Bravo, and Nanda Wijermans, editors, *Advances in Social Simulation*, pages 15–25, Cham, 2020. Springer International Publishing.

Designing for Presence: Embodied Interaction in Computer-Mediated Realities

Holger Regenbrecht

University of Otago, Dunedin, New Zealand

holger.regenbrecht@otago.ac.nz

Thomas Schubert

University of Oslo, Norway

t.w.schubert@psykologi.uio.no

Abstract

There is only embodied interaction in computer-mediated realities. We use our bodies to sense, interact with, process, and remember environments. Effective computer-mediated realities, i.e. mixed and virtual realities, depend on the user's feeling of being part of those environments. We are explaining our approach towards a theoretical model for this sense of presence. Based on techno-psychological, philosophical, and socio-psychological models we are presenting a theory for the understanding of and design for such mediated realities. We emphasize the role of the human body as a sensing and interacting instrument and argue that presence is achieved by self-controlled navigation, object interaction, and subject interaction. Presence therefore derives from body related possibilities of interaction with the environment. We demonstrate this through the definition of a theoretical presence model for embodied interaction in computer-mediated realities.

This article is dedicated to Martin K.Purvis—one of the few modern renaissance scholars who is not only well educated and well read, but who continues to inspire young and curious minds, We hope that our thoughts presented here will inspire and entertain Martin as well as other readers with interests in the crossing boundaries of technology, psychology, and philosophy.

1 Introduction

Computer-mediated realities, i.e. three-dimensional worlds that can be explored interactively in real-time, currently experience a revival. These worlds range from augmented reality (AR), where physical (real) reality is enriched with spatially registered virtual objects, via augmented virtuality (AV), where virtual worlds are enriched with real objects to virtual reality (VR), where the entire experience happens within an immersive environment, leaving out any signs of physical reality (if possible and desirable). The virtuality continuum [17] labels the range between AR and AV as mixed reality (MR) while computer-mediated realities also include VR [22].

Such computer-mediated realities (CMR) should evoke a feeling of being present for their users. To feel present means that the user feels that he/she is part of the mediated environment, accepting it as reality or world. Indeed, it seems that these computer-mediated environments aim at, benefit from, or even require a sense of presence in order to be used to their full extent. Only with a sense of presence, users will apply their full range of intuitive and heuristic reasoning and transfer knowledge from other environments in which they feel present—reality itself. In other words, computer-mediated realities can be understood as interactive, three-dimensional worlds which are considered as realities by users.

A rich and diverse field of psychological research has been conducted with the goal to understand and develop the relationship between individuals and technical environments. Within it, there exists a body of research on the sense of presence in virtual environments that has a long tradition (e.g. [1, 27, 30, 14, 24, 25, 11]). Notably, a lot of it was discussed in the 1990's during the first wave of VR and AR technology development. Since then, the technology has recently seen another breakthrough with the advent of affordable and high quality display, graphics, and sensing technology.

We propose to investigate the sense of presence in CMRs by way of considering techno-psychological, philosophical and socio-psychological sources. We try to integrate knowledge from those sources to develop a framework model on CMR presence and to derive some design propositions for CMR. Presence,defined as "being in" the CMR, raises questions about "being in worlds" in general. How do body and mind relate to the surrounding environment?

Philosophically, those questions can be traced back to the Sophists and lead to a rationalism dilemma: We want to consider user (subject) and environment as one, non-divisible unit, because this is what being in an environment essentially is. On the other hand, we have to consider these two realms separately to be able to explain and communicate. Therefore, in the following we will go back and forth between these two views: an integrated, closely coupled human-environment view and a separation of the two realms where they are loosely coupled.

In both types of views, computer-mediated realities are unlike other media or technologies before. We were either present in the present (expressed with the present tense) and called this reality, or we used media, arts, and culture to be present in the past or future or imaginary, e.g. by way of storytelling or moviemaking. CMRs fold past, present, and future into one, direct experience of a feeling of being present in that computer-mediated reality—also called the sense of presence.

Briefly, our argument is the following: Human cognition is in the service of action in the world. Successful action requires being informed by knowledge about what interactions are possible in the HERE AND NOW, how interactions in the PAST went, and predictions about outcomes of actions in the FUTURE. Only possibilities in the HERE AND NOW are the immediate, direct outcome of perception; psychologists have called this the outcome of projectable information. Information about the PAST and predictions about the FUTURE are contributed by our minds; this has been called non-projectable information [7].

Because non-projectable information about PAST and FUTURE is so important, we share them with each other, being the social species that we are. For that purpose, we have created language and other media, and we spend a large part of our childhood learning to understand these media properly. Understanding means being able to build mental representations that allow us to act.

Indeed, we are so good at mentally simulating PAST or POSSIBLE worlds that sometimes they start to "feel" real, or that we feel present in them (e.g., in our imagined worlds, or in worlds that we are told about in books, in movies, etc.). For a long time, philosophers and psychologists have studied how humans build these mental representations from media and how they use them to enhance their interactions with the real environment.

With CMR, what happens now is that media fully enter the world of projectable information, because the presentation of media content is now indexed to our body's movement. Thus, CMR now conquer the HERE AND NOW. However, that does not eliminate established media representations of other types, because they are amazingly efficient, and we already know how to use them. As a result, media representations of all kinds co-exist in CMR.

This analysis offers a framework for anyone trying to understand and create CMRs by adapting relevant philosophical foundations combined with techno-psychological and socio-psychological research. First, the CMRs' simulation of the HERE AND NOW in and by itself has to work as intended; psychological knowledge on how that works helps us to do that. Second, psychological research has investigated how understanding of the HERE AND NOW is integrated with information provided by media. Because CMRs contain both, this work is directly relevant to CMRs as well.

A sense of presence develops in three different, sometimes co-occurring types: spatial presence, social presence, and co-presence. Spatial presence, also known as the sense of "being there" breaks down to the components of the defining spatial presence, involvement, and realism [24]. Social presence or the sense of "being together" and co-presence, as the sense of (spatially) "being next to one another" are of high importance for group experiences of CMR, but are beyond the scope of this paper. Spatial presence, or simply presence, can and should be considered as a feeling [25]. Therefore, we have to view presence as caused by unconscious processes, as being immediate, in a philosophically loosely coupled view as having no truth value (can't be judged), and as having levels of intensity, amongst other aspects. Again, presence is the (individual) feeling of experiencing reality in the HERE AND NOW.

We are going to argue that presence develops by and through embodiment, which we break down into embodied interaction and embodied cognition. The embodied interaction aspect considers the human body as the sensing and interacting instrument in CMR while the embodied cognition aspects considers the conceptualization of reality.

The term "embodied interaction" was prominently introduced to the literature on human-computer interfaces by Paul Dourish [2]; this was often misread as being equivalent to tangible computing [3]. Dourish's notion of embodied interaction is one on the relationship of technical and social considerations. Hence, it is rather about "an embodied account of interaction for traditional user interface design and analysis" [3, p. 2:2]. We build on this notion, but focus on the relationship of technical and psychological and socio-psychological considerations. By doing this, we merge Dourish's view with Kirsh's [16] view on embodied cognition to again decompose it into the tangible, perceptual, and instrumental aspects and the conceptualization aspects. Kirsh's embodied cognition theory proposal is grounded in four ideas: (1) that interaction with tools influences perception and thinking, (2) that thinking is embodied, (3) that doing contributes more to knowing than seeing, and (4) that sometimes we literally think with things [16, p. 3:1]. If we combine both authors' views, we end up with an unsatisfactory ambiguity in terminology. In a philosophically loosely coupled way, we propose to separate embodiment into embodied interaction and embodied cognition (although we are aware of the rationalist dilemma within this).

In the remainder of this paper, we use Heidegger's concept of being in this world as a philosophical scaffolding to explain interaction and presence in CMR. Of particular interest are the aspects of Zuhandendem and Vorhandenem and the transition from Sein to Dasein. Both aspects can be seen as a strong account for the essence of embodiment and presence for CMR.

We then discuss three models of perception and how they relate to presence, leading to the conclusions that human cognition is embodied, that humans act on projectable information, and that past and future (non-projectable) experiences are meshed into affordances shaped by the format of the medium, here CMR. This discussion is followed by considerations on how CMRs are mentally constructed and presents an argument why CMRs are neither equivalent nor arbitrary interpretations of worlds. We shed a different light on mental models as co-existing and partial (analog) representations to be considered for designing CMRs. Finally, we present a set of propositions summarizing our model in a more tangible way for designing for presence.

For the composition of our model, we start of by considering ways of understanding perception which are predominantly used in theory and application of interaction design and human-computer interaction research.

2 Philosophical Scaffolding: M. Heidegger

The work of German philosopher Martin Heidegger can be divided into two phases: (1) his earlier work considered the concept of being and being in this world (In-der-Welt-Sein) and (2) his later work dealt with the relationship between men and technique (and technologies). Both aspects are of high relevance to presence in computer-mediated realities and therefore will be examined here [1]. His philosophy developed complex concepts that he typically referred to with newly coined or abstracted German terms.

Heidegger asks what the definition of being (Sein) is and what it differentiates from what is, or the "entity" (das Seiende), some specific thing that is. The entity (Seiende) can be found in all material objects,

---

[1] We are aware of the complicated side of Heidegger. During the rise of fascism in Germany, he evidently took the side of fascist movement and supported discrimination of Jewish colleagues. We believe that much of his work can be interpreted despite these condemnable behaviours.

in abstract measures, in imaginations, in subjects (people), or even in the absolute (the divine). Building on this distinction, he introduced the term Dasein, the human form of being (Sein). The essence of Dasein lies in its very existence [12, p. 42]. Dasein is achieved by understanding, exploring, or in Heidegger's terms, grasping (begreifen, which is the German term for grasping, both in its literal and abstract sense). Through grasping (begreifen), we move from mere being-in-the-world (In-der-Welt-Sein) to Dasein. When merely being in the world (In-der-Welt-Sein) we can't distinguish between the self and the world, even self-doubt happens within the world. Grasping (begreifen) is strongly linked to interaction and the situation we are (inter-)acting in.

Heidegger also introduced the term thrownness (Geworfenheit), loosely translated as "to be thrown". We are thrown into this world (whether we like it or not) and are made to act! Humans are in a constant state of thrownness, in a constant state of acting.

The world in turn, is not simply there; the world reveals itself by Zuhandenem—tools or objects with a purpose or being ready-to-hand. To illustrate this thought Heidegger used the example of a hammer: The hammer is out there in the world to be used as a hammer. This makes it part of the (current, individual) environment, it is ready to hand (zuhanden). During the process of hammering, the hammer as something ready to hand disappears, it simply serves its purpose and is present-at-hand (vorhanden) (simply there, but with this "invisible"). Only if the hammer fails its function (e.g. brakes) it will become apparent again, and with this ready to hand (zuhanden). Therefore, in Heidegger's world it is reasonable to explain and name things and matters in terms of their readiness-to-hand (Zuhandenheit, being zuhanden) and not simply in terms of their apparent (surface) properties.

To understand presence in CMR, we can use Heidegger's concepts in the following ways: Virtual worlds are embedded in the real world to different degrees, ranging from augmented reality to immersive virtual reality. For a user, there must be (enough) clues or hints to realize or accept the presented CMR as a world. It is not sufficient to just provide a being-in-the-virtual-world (in-der-virtuellen-welt-sein), rather, the CMR has to provide a set of clues and objects to interact with, of Zuhandenem. Only then, the user can turn the ready-to-hand (zuhanden) objects in the world into present-at-hand (vorhanden) objects by interacting with the world and with this coming into a state of Dasein. The being-in-the-world becomes Dasein, becomes presence. The (technical) CMR apparatus provides a situation of thrownness (Geworfenheit), which makes the user act in.

The Heideggerian concepts adapted so far stemmed from his earlier period of work. Later he explored humans' interactions with the technological world they created. His notion of the Ge-stell can serve as guidelines for understanding technological systems. We will use it here to consider the technical apparatus of a CMR system, i.e. the technology surrounding the user as well as the (3D) content and possible interactions. Ge-stell can be loosely translated into the frame, the scaffolding, the stage, or the background that is used to set the CMR scene. It serves to reveal (entbergen) the hidden (verborgene). It includes all forms of interaction like production or presentation. For CMR design that means, that we should keep the user away from consciously working in and for the Ge-stell. He/she should rather interpret (entbergen). In a simplified way, we could even say that the focus is on content, not technology.

Sein and Dasein have to be seen in relation to the self—the interplay of self and environment leads to presence.

Presence develops in the Ge-stell through action. Presence can only develop if the environment is recognized and accepted as a world (reality) by the user (self). CMR technology is a necessary, but not sufficient prerequisite for presence.

3 The Flow, the Dam, And the Kayak: Three ways to understand perception

Psychological theorizing has proposed at least three classic ways of how to understand what happens during perception. These three classics will inform our discussion because they obviously apply to the perception of virtual environments, but they will also help us to better understand what virtual environments actually do. In short, the three ways are the Gibsonian view (the environment affords itself to us), the Fodorian view (we recognize objects and their properties), and the embodied way (perception is embodied action). Let us explain each of these classic views in turn.

3.1 The Gibsonian View: There is only flow

J.J. Gibson [9] developed a model of perception, which assumes that we pick up (apparent) properties from our environment, defined by an optical flow array. The perceiver selects (picks) those properties from a rich information environment that are of immediate relevance. Those properties form an ad-hoc individual environment, which is not based on mental representations but rather on directly perceivable properties.

The environment is both, the object of perception as well as the perceptual information. The perceiver and the environment form one, tightly coupled unit. Information from the environment is to be considered of relevance if they require action, becoming so-called ecological objects. Gibson introduced the term affordances to describe those action-relevant properties of objects. The environment consists of affordances, and only those are relevant.

An ontological consequence of such a thought is the necessary interplay between environment and perception to describe the nature of existence of organisms, here humans. The perceiver influences the environment—the environment influences the user. Distinguishing between subject and object is almost impossible (the rationalism dilemma). Hence, the only measure for perception is effective action in the environment. According to Gibson, all other constructs, like mental models, are redundant.

The claimed redundancy of models sparks two essential questions: (1) How can effective acting be defined and measured? Applying Gibson's framework to CMRs, Zahorik & Jenison [31] defined effective action as any acting which finds its equivalence in the real world. I.e. if we can interact with a CMR in the same way as in the real world then we would call this effective. Obviously, CMRs don't have to follow the same laws as the real world (e.g. gravity), so this might invalidate this argument. Nevertheless, Zahorik & Jenison argue that to date all development (cultural, social, psychological) happened in the real world, so was learned only there. (2) Aren't there objective indicators, like time needed to complete a task to suffice the requirement of measurability?

3.2 The Cognitivist Dam: There is flow, and then there is knowledge

Fodor & Pylishin [7] criticized Gibson's model in many ways. Of particular relevance to us is their argument that we need to differentiate ecological (affordances) from non-ecological objects.

> "..., if any property can count as an invariant, and if any psychological process can count as the pickup of an invariant, then the identification of perception with the pickup of invariants excludes nothing." (p. 141)

> "Our argument will be that (a) the prototypical perceptual relations (seeing, hearing, tasting, etc.) are extensional [...] (b) whereas, on the contrary, most other prototypical cognitive relations (believing, expecting, thinking about, seeing as, etc.) are intentional; and (c) the main work that the mental representation construct does in cognitive theory is to provide a basis for explaining the intentionality of cognitive relations." (p. 188)

To solve this problem, they offer the concept of projectable and non-projectable properties of perception. Projectable properties are those which can be directly picked up from the environment, like Gibson's optical flow array. Non-projectable properties, on the other hand, require additional information from the user's knowledge. They are not obvious, apparent or directly perceivable.

Take as an example a bottle of soda. The bottle is designed and manufactured to afford the actions of being picked up with one hand, being tilted, and pouring the liquid—if you wish, directly into your mouth, with just the right aperture to fit it. All that information is projectable but comes into being in conjunction with a body able to pick up, tilt, and drink. On the other hand, the drinkability and digestibility of the liquid of a certain consistency is most likely not hard-coded into our genetic makeup but learned as a fact, and thus not projectable.

3.3 Navigating the flow: Embodied action

There are major differences in the views of Gibson and Fodor & Pylishin. Nevertheless, we can extract some valuable, pragmatic thoughts here:

1. Perception in CMR is based on an action-perception loop. There is no action-independent perception.

2. There are projectable and non-projectable properties in the CMR.

Varela, Thomson, and Rosch [28] attempted to mediate between the different object-subject viewpoints, which are reflected in their extremes by rationalism/positivism and idealism/solipsism, with their proposed Embodied Mind approach. They developed a cognitive science theory inspired by Buddhist philosophy and embodied perception. Their approach can be summarized as an:

> "... emphasis on mutual specification ... [that] enables us to negotiate a middle path between the Scylla of cognition as the recovery of a pregiven outer world (realism) and the Charybdis of cognition as the projection of a pregiven inner world (idealism). ... Our intention is to bypass entirely this logical geography of inner versus outer by studying cognition not as recovery or projection but as embodied action." (p.209)

These two views, also referred to as the hen-egg problem, address the question of objectivity. On one side, the hen position, there is an objective world which can be represented or imagined by the perceiver. On the other side, the egg position, the world is entirely constructed by the perceiver. In both cases we talk about representations, either constructed externally or internally.

The combination of both representations, however, introduces the concept of embodied action. Cognition cannot be decoupled from the biological, psychological, or cultural background of the perceiver and experiences resulting from sensory and motor interaction. Varela et al. developed a model of enaction: Perception is always perception-driven acting and acting is based on cognitive structures made of sensorimotor patterns. The cognitive model is based on the sensorimotor experiences of the self.

This model of Varela et al. has contributed to the field of embodied cognition, which has emerged in the last two decades. For instance, cognitive scientist Arthur M. Glenberg [10] proposed a theory of memory building upon embodied cognition concepts. His approach focused on both perception of real actual environments, language, and their interaction.

Like Varela and Gibson, Glenberg considered perception as an integral sensorimotor process; he sees the body as the central element of this perception loop. On one hand, the body is the main reference frame for all actions; on the other hand, it is also the instrument for sensorimotor perception: acting and perceiving. Experiences made this way are stored as embodied experiences in memory.

While Glenberg also recognizes projectable and non-projectable properties he claims that objects can only be perceived if there are possible actions. Importantly, and in line with the general notion of embodied cognition, Glenberg argued that non-projectable properties are always stored as embodied experiences, i.e. they are not independent from the situation in which they have been (sensomotorically) generated.

A continuous meshing of projectable and non-projectable properties leads to implicit or indirect memory.

"When we are walking the path home, we do not need to
consciously recall which way to turn at each intersection";
[10, p. 10]

We conceptualize a meshed set of patterns of action.

Using dangerous, everyday situations as examples, Glenberg
demonstrates that normally the projectable properties dominate. They

"...may not require any sort of representation of the environment and they may not require memory."(p. 4).

He calls this concept clamping. Non-projectable properties are
clamped to develop different, situation-specific conceptualizations resulting in different possible meanings for a given situation:

"...the world is conceptualized (in part) as patterns of possible bodily interactions, that is, how we can move our
hands and fingers, our legs and bodies, our eyes and ears,
to deal with the world that presents itself" (p. 3).

He emphasizes the assignment of meaning by possible actions.

The process of meshing projectable and non-projectable information is exemplified by understanding language that refers to a situation. E.g., imagine trying to understand instructions of how to find
your way to the bus stop. In order to find your way successfully,
you have to generate mental representations of possible actions, mesh
them with your current state, and then execute and update them
based on your locomotion.

However, sometimes it is necessary to suppress certain perceptions.
This is in particular necessary if there are contradicting or incoherent sensations and perceptions. Suppression implies mental load. If
you try to understand instructions that do not refer to your current
situation but some future state, you need to partly or completely suppress representations based on current projectable information. This
is what you do when you read novels or watch TV.

Conceptualizations are constantly changing and form trajectories.

"Here is the proposal for updating memory: memory is updated automatically (that is, without intention) whenever there is a change in conceptualization (mesh)." (p. 7).

Later we refer back to those trajectories as feelings of memory to mesh new conceptualizations. Those trajectories are embodied, i.e. they "describe" the memorized situation in relation to own actions and therefore are easily applicable to the current situation.

"...memory is embodied by encoding meshed (i.e. integrated by virtue of their analogical shapes) sets of patterns of action. How the patterns combine is constrained by how our bodies work. A meshed set of patterns correspond to a conceptualization."(p. 3).

3.4 Summary

To sum up: Cognitive scientists have grappled for decades with the ontological status of the information that humans use to adapt their behavior. Is it out there, and we take it in, implying a difference between us and the environment, or is there never anything in us, but we simply interact? A viable (and informative!) solution seems to be to acknowledge that:

1. Human cognition is always in the service of action, and action always requires a body.

2. Humans act on the basis of information picked up directly from affordances in the environment. Such affordances are defined jointly by the environment and the body. We can call this type of information projectable information.

3. However, action is also shaped by information that is either remembered from past experiences, or information that is generated in simulations about possible future environments. This type of information can be called non-projectable. Humans mesh such information with information from affordances picked

up from flow. The meshing is eased by the format in which these mental representations (memories and simulations) are run, namely in an embodied, modal, format—the same format that the direct perception of the environment based on affordances uses as well.

4 Interpretation of Worlds and a Model for Prescence

As indicated earlier, computer-mediated realities form a subset of possible alternative realities humans might construct. Even if we reduce this set of worlds to digitally constructed ones we might struggle to formulate what can and what should be included in our considerations. From a technological point of view, we might define CMR as: (1) computer generated, (2) three-dimensional, (3) interactive with real-time feedback, and (4) immersive (aiming for a user's sense of presence in that reality). From a content provision point of view, environments can be offered which reflect as close as possible real-world (physical world) situations, as for instance needed for VR or AR supported design reviews of to-be-manufactured goods. Alternatively, artistic environments might be created with properties beyond what we know from the real world (within limits).

Now, let us apply the insights from cognitive science on how humans perceive and interact with CMRs. Traditional media, such as books, but even spoken language, provide us with non-projectable information that, once processed and mentally represented, can be meshed with projectable information from the environment. This is how we learn that the red end of the compass needle points north (where we have to walk to get home), or that milk is bad for us if we are lactose intolerant, and how we hopefully act adaptively based on this knowledge. The information that we acquire this way can be about the past or about possible futures. Interestingly, any non-projectable information conveyed in a medium has of course to hitchhike on actual perception of the real environment and thus projectable information: the letters on this page are made of actual atoms on paper or a screen.

89

The distinction of projectable and non-projectable information gives us an understanding of what CMR does in terms of perception: The truly amazing achievement of CMR technology is that it goes beyond providing non-projectable information that we still need to mesh. Instead, it is a medium that provides projectable information. It shares this feat only with a few other media forms that usually are not nearly as efficient, such as (participatory) theatre, role play, or, in few rare moments, movies.

To appreciate the difference, consider how you would perceive and apprehend a photo: First of all, the projectable information provided is that the photo is displayed on some kind of surface, and that surface can interact with your real body. In addition, you can construct a spatial mental representation of the environment depicted in the photo, which is an acquired skill—but this is not based on projectable properties directly—it requires interpretation and additional cognitive processing. Only after doing this, you can suppress your real environment and mentally place yourself in the depicted environment. Of course, all these processes are happening rapidly and efficiently in adults, and are mentally impenetrable. A CMR allows the perceiver to skip these steps.

Nevertheless, CMRs do not only convey projectable information. Because it can simulate projectable information, it can also simulate any other medium (with hopefully enough resolution!).

Varela et al.'s and Glenberg's notions have been the basis of our own approach to develop a model for presence in CMR. Human actors in a CMR receive projectable information from the real environment (e.g., the weight of the head-mounted display on their head and the floor they walk on), projectable information from the CMR that changes in accordance with movements and other actions (e.g., speech), and non-projectable information from the CMR (e.g., labels). They may even receive additional non-projectable information from outside the CMR, for instance if a VR session is accompanied by instructions what to do in the virtual environment. Accordingly, spatial presence, when defined as a subjective experience, develops when control of the own body is seen as a possible action in the CMR. Presence can be increased if more meaningful possibilities of interactions are offered and appropriate conceptualizations can be made [20].

Also, the possible actions with objects in the environment lead to a better understanding of the CMR. If those actions are perceived as body-related, then this will lead to presence within the environment. Body-related stimuli increase presence.

The need to suppress incoherent stimuli leads to the opposite effect: presence will be decreased or will break.

4.1 Mental Models and Metaphors for CMR Design

If we understand embodiment as embodied interaction and embodied cognition, which cognitive models could be applied to inform CMR design? How do users of CMR represent the surrounding environment? Is it rather a symbolic-abstract representation, also called propositional, or is it a rather analogue or in laymen's terms representational model?

The nature of CMR, their spatial extents and their appearance suggest themselves to analogue models. Based on theories developed by Johnson-Laird [15], Schnotz [23] developed an approach of mental models as representations of actual situations. Mental models are used to build constructions about the matter, structure, and function of "worlds".

Mental models are based on three principles:

1. The principle of computability: Mental models, and the machinery for constructing and interpreting them, are computable. [...]

2. The principle of finitism: A mental model must be finite in size and cannot directly represent an infinite domain. [...]

3. The principle of constructivism: A mental model is constructed from tokens arranged in a particular structure to represent a state of affairs. [15, p. 398]

Those principles show that mental models (a) are interpreted constructions, (b) those constructions are simplified to a degree that they are "computable" (also "principle of economy of models"), and (c) that it is impossible to transfer (objective) realities to internal representations.

Mental models are constructed before the background of individual experiences.

> "...our view of the world is causally dependent both on the way the world is and on the way we are." (p. 402).

Possible concepts which can be represented in mental models are categorized into: time, space, possibility, permissibility, causation, and intention. Using those concepts, mental models organize structures for possibilities on semantic operators. Perceptual processes normally build a three-dimensional (kinematic) model which relates volumina of objects into an object-centric coordinate system, hence the suitability of mental models as models to explain the perception of and interaction with CMRs.

Purely propositional models, based on for instance words or symbols, are rejected by Johnson-Laird. He refers to the behavior of little children or to the difficulty in distinguishing between object and belief. Important is the representation:

> "What Meister Eckart believed about God could not be expressed in words ..." (p. 432).

Johnson-Laird gives us a cognitive theory which allows to develop categories of spatial structures and with this the relation of the self to the surrounding environment.

4.2 Plausibility of CMR

In the style of Eco (1990/1995) we are distinguishing between categorizations of worlds as possible/impossible and as imaginable/unimaginable. Normally, and if seen from the viewpoint of the designer of a CMR, imaginable and possible worlds are the norm. We want to create augmentations for the real world which (almost) seamlessly blend in with reality or we want to create virtual worlds, even if they are immersive and therefore (partially) decoupled from real stimuli, with properties known from real world experiences. If those worlds are plausible, i.e. possible and imaginable, we call them effective.

There are worlds which are imaginable but normally impossible, like speaking animals in fairy tales. In CMR, these worlds require some sort of willingness of the user to accept, at least for a certain while, the lack of plausibility in those situations.

Possible and impossible, imaginable worlds can be communicated by writing about them, for instance, those worlds are of lesser interest for CMRs.

Imaginable (possible and impossible) CMRs can be interpreted differently, depending on one's perspective. On one hand there is the intention of the designer of the world who makes an offer for the recipient to interpret the world. Eco talked about a model reader (because he wrote mainly about texts), hence a fictive person or group of persons, which functions as a proxy during the design process. In interaction design those persons are sometimes referred to as personas. For CMR we might call this fictive person, our imagined user, proxy actor. This proxy actor, on the other hand, explicitly or implicitly forms a picture of the creator of the CMR; what does this world expect from me?; What is the intention? This includes not only the (imagined) person of the designer but also includes algorithms, the "machine" as such, or the "artificial intelligence" of the CMR - the proxy author.

Norman [18] discussed in a practical way what creators (authors) and recipients (actors) might have in their minds when developing and using CMRs.

The author's design goal is to offer a CMR which expresses a certain form (appearance), a (narrative) plot and possibilities for interaction. Using the concept of mental models the following constructs develop:

- MMauthor1 (real world): mental model of the author about the environment the proxy actor lives and acts in. This is characterized by "nature and nurture" factors.

- MMauthor2 (CMR): mental model of author about ten to-be-created CMR. This model often changes many times during the

93

development process.

- MMauthor3 (actor): mental model of the author about the proxy actor; can be a real person group, or a fictive user (e.g. persona)

- MMauthor4 (MMactor (CMR)): the mental model of the author about the actor's mental model of the CMR. This model addresses the question how the actor might construct a model about the developed CMR.

- MMauthor5 (MMactor (author)): the mental model of the author about the mental model of the actor about him/herself.

Whether and how often and how deeply authors think about this is unknown, also whether further considerations or recursions happen in the form of MMauthor (MMactor (MMauthor (...))).

The author tries to provide the proxy actor with a set of stimuli to satisfy his mental model about the to-be created CMR and the user's mental models about it. An equivalent interpretation, i.e. a 100% match of those mental models is impossible.

From the user's (actor's) point of view there is a corresponding set of mental models:

- MMactor1 (real world): the mental model of the user about his/her real environment, including nature/nurture background.

- MMactor2 (CMR): the mental model of the user about the CMR at hand. This model changes during interaction, it would therefore be helpful to (at least) break this mental model into three phases: before, during, and after use/experience.

All other possible mental model relationships can be developed in a similar way as for the author's view. Even if an equivalent interpretation by the author and actor of the CMR is impossible, that does not mean that there is an arbitrary number of interpretations, and neither is there a unique interpretation. There are interpretations which are inadequate. In Eco's words relating to texts: "...any act of interpretation is a dialectic between openness and form, initiative on the part of the interpreter and contextual pressure" (p.21). Hence, the interpretation of a CMR happens within the limits of equivalent and arbitrary interpretation, but excluding the extremes.

We now have a scope for representations and interpretations of CMR. Let's now "throw" a user into our CMR environment. Will that user develop a feeling of presence, i.e. the sense of being part of that CMR?

5 Propositions

We are presenting a set of propositions to be considered when designing and analyzing computer-mediated realities.

The sense of presence, in particular spatial presence, can best be explained with mental models and not with propositional approaches. Propositional elements might be present within mental models though. Presence results from the user's effort to imagine and to construct a world including her/his relations to this world. This requires (a) a willingness to perceive an alternative (computer-mediated) world and (b) the willingness to put in this effort.

Presence only develops if, and only if the CMR is perceived as such. There have to be stimuli which allow for an appropriate interpretation. The design of the CMR determines the possible room of interpretations, which is not arbitrary. This does not mean, that there aren't any interpretations which fall outside of the intentions of the designer. Presence develops in a biological, psychological, and cultural context of a user. It can't be considered outside the background and experience of the user. Those experiences are mainly based on real/physical world interactions and perceptions. These days this includes the interaction with computing artifacts.

Interactions can be of different types: imagined, planned, planned and executed or simply executed.

The to-be-perceived environment consists of projectable and non-projectable properties. Projectable properties are perceived sensorial in a direct manner and do not require cognitive effort. Non-projectable properties are amended with experiences from memory, resulting in changing conceptualizations of the situation at hand. CMR have the potential to fold past and future into present and with this into projectable properties.

Objects within the CMR are either relevant for orientation or for action. Orientation-relevant objects are perceived in the "being a" aspect of an object, don't require (or only little) mental resources and therefore have projectable properties. Action-relevant objects are perceived as objects of possible actions - they can have projectable and non-projectable properties. One and the same object can be interpreted by the users as orientation- or action-relevant depending on the situation at hand. The current goal of action determines the mode of interpretation.

There should be more projectable than non-projectable properties in a CMR to maximize the likelihood of presence.

The more non-projectable properties one finds in a CMR the more interaction is required to conceptualize the environment. Von Foerster's [8] 4D cube experiment might serve as an example here: When only watching other users interacting with the a stereoscopic hypercube (a mathematical 4D→3D →2 × 2D projection) the environment appeared to have almost only non-projectable properties. Only when users interact with the system for a certain while the environment could be conceptualized (grasped).

Presence develops through interaction with the CMR. There are three forms of interaction (cf. Regenbrecht, 2000):

1. self-controlled movement through the environment

2. interaction with objects

3. interaction with subjects (communication and cooperation)

Presence develops through an action-perception loop or sensory-motoric perception. While action and perception form one concept they have to be separated for (rationalistic) model building here.

The act of acting in CMR is body-related, it is embodied! Objects of the CMR are interpreted in relation to the own body. That means that

(a) all Cartesian variables (like distance, size, or position in space) are put into relation to the body (with all their projectable and non-projectable properties) and (b) all possible actions are related to the body as an interaction instrument.

Action-relevant objects in a virtual environment do increase the sense of presence if they are presented in a body-conform way. Orientation-relevant objects increase the sense of presence if they are presented as part of the experience context (also embodied) of the user in their "being a" aspect and therefore immediately projectable.

The suppression of incoherent stimuli between the different domains of a CMR, i.e. real, augmented, and virtual, demands mental resources. If a user can't free up those resources transparently the sense of presence will decrease or break. This leads to two ramifications: (1) the easier the (intended) conceptualization can be achieved the less has to be suppressed and (2) the fewer incoherent stimuli from other domains are perceived the easier the conceptualization.

The meaning of the CMR as a whole (not necessarily of its single elements) is determined by the possible, body-related actions of the user. Those actions derive from possible patterns of actions (intention of the designer of the CMR) and the patterns of actions of the user. Conceptualizations lead to possible actions in the CMR.

The more relevant a CMR is for a user the higher the sense of presence (including non-spatial presence). Spatial presence develops through body-related possibilities for interaction within the CMR.

There is only embodied interaction in computer-mediated realities.

Acknowledgements

We would like to thank the members of igroup.org for all the fruitful, and sometimes very emotional and time-consuming discussions we've had about life, the universe, presence, and everything: Martin Kohlhaas, Frank Friedmann, Jakob Beetz, Thore Schmidt-Tjarksen, Ernst
Kruijff, Jan Springer, Tobias Hofmann†, and all the other igroup people!We would also like to thank Otago's HCI group, in particular

Tobias Langlotz, Jonny Collins, and Mike Goodwin†.

References

[1] Barfield, W., & Weghorst, S. (1993). The Sense of Presence Within Virtual Environments: A Conceptual Framework Human-Computer Interaction. Proceedings of the HCI International '93. Amsterdam-London-New York-Tokyo: Elsevier.

[2] Dourish, P. (2001). Where The Action Is: The Foundations of Embodied Interaction. MIT Press.

[3] Dourish, P. (2013). Epilogue: Where the action was, wasn't, should have been, and might yet be. ACM Transactions on Computer-Human Interaction (TOCHI), 20(1), 2.

[4] Dreyfus, H.L., & Dreyfus, S.E. (1988). Making a Mind Versus Modeling the Brain: Artificial Intelligence Back at a

[5] Branchpoint. In Graubard (ed.). The Artificial Intelligence Debate. Cambridge, 15-43.

[6] Eco,U. (1995). Die Grenzen der Interpretation [The Limits of Interpretation]. München: dtv wissenschaft.

[7] Fodor, J.A., & Pylyshyn, Z.W. (1981). How direct is visual perception?: Some reflections on Gibson's "Ecological Approach". Cognition 9(1981). 139-196.

[8] Foerster, H.v. (1992). Entdecken oder Erfinden. Wie läßt sich Verstehen verstehen? [How can understanding be understood?]. In Glaserfeld, E.v. (Hrsg.) Einführung in den Konstruktivismus [introduction into constructivism]. München: Piper, 41-88.

[9] Gibson, J.J. (1979). The ecological approach to visual perception. Boston: Houghton Mifflin.

[10] Glenberg, A. M. (1997). What memory is for. Behavioral and Brain Sciences (1997) 20:1, 1-55.

[11] Grimshaw, M. (editor) (2014). The Oxford Handbook of Virtuality (Oxford Handbooks). Oxford University Press.

[12] Heidegger, M. (1957). Sein und Zeit [being and time]. Tübingen.

[13] Heidegger, M. (1962 / 91). Die Technik und die Kehre [technic and turn]. Pfufflingen: Günther Neske

[14] IJsselsteijn, W.A., De Ridder, H., Freeman, J., and Avons, S.E. (2000). Presence: Concept, determinants and measurement. Proceedings of the

SPIE , 3959, 520-529. January. 2000.

[15] Johnson-Laird, P.N. (1983). Mental Models. Cambridge, MA: Harvard University Press.

[16] Kirsh, D. (2013). Embodied cognition and the magical future of interaction design. ACM Transactions on Computer-Human Interaction (TOCHI), 20(1), 3.

[17] Milgram, P., Takemura, H., Utsumi, A., Kishino, F., Augmented Reality: A Class of Displays on the Reality-Virtuality Continuum. (1994) Proceedings of Telemanipulator and Telepresence Technologies. 1994. SPIE Vol. 2351, 282-292. Program: Part I - Vol. 145 (July 27–27, 2003). ACM Press, New York, NY, 4. DOI:http://dx.doi.org/99.9999/woot07-S422

[18] Norman, D. A., & Draper, S. W. (eds.) (1986). User Centered System Design: New Perspectives on Human-Computer Interaction. Hillsdale,NJ: Lawrence Erlbaum Associates, Publishers.

[19] Regenbrecht, H. (1999). Faktoren für Präsenz in virtueller Architektur [Factors for the sense of presence in virtual architecture]. Unpublished doctoral thesis (Dissertation). Bauhaus University Weimar, Germany.

[20] Regenbrecht, H. & Schubert, T. (2002a). Real and Illusory Interactions Enhance Presence in Virtual Environments. Presence: Teleoperators and virtual environments, 11(4), MIT Press, Cambridge/MA, USA. 425-434.

[21] Regenbrecht, H. & Schubert, T. (2002b). Measuring Presence in Augmented Reality Environments: Design and a First Test of a Questionnaire. Proceedings of the Fifth Annual International Workshop Presence 2002, Porto, Portugal - October 9-11.

[22] Regenbrecht, H., Hoermann, S., Ott, C. Mueller, L., & Franz, E. (2014). Manipulating the Experience of Reality for Rehabilitation Applications. Proceedings of the IEEE 102(2), February 2014, 170-184.

[23] Schnotz, W. (1994). Aufbau von Wissensstrukturen: Untersuchungen zur Kohärenzbildung beim Wissenserwerb mit Texten [building knowledge structures during knowledge acquisition with texts]. Weinheim: Beltz, Psychologie-Verl.-Union.

[24] Schubert, T., Friedmann, F., & Regenbrecht, H. (2001). The experience of presence: Factor analytic insights. Presence: Teleoperators and virtual environments, 10(3), MIT Press, Cambridge/MA, USA. 266-281. Schubert, T. W. (2009). A new conception of spatial presence: once again, with feeling. Communication Theory, 19(2), 161-187.

[25] Schubert, T. W. (2009). A new conception of spatial presence: once again, with feeling. Communication Theory, 19(2), 161-187.

[26] Shannon, C. & Weaver, W. (1962). The mathematical theory of communication. Urbana, IL: University of Illinois Press.

[27] Slater, M., Usoh, M., & Steed, A. (1994). Depth of Presence in Virtual Environments. Presence: Teleoperators and Virtual Environments, 3(2), 130-144.

[28] Varela, F.J., Thompson, E., & Rosch, E. (1991). The Embodied Mind. Cambridge, MA.: MIT Press.

[29] Winograd, T., & Flores, F. (1992). Erkenntnis Maschinen Verstehen [Understanding Computers and Cognition]. Berlin: Rotbuch Verlag.

[30] Witmer, B.G., & Singer, M. (1998). Measuring Presence in Virtual Environments: A Presence Questionnaire. Presence: Teleoperators and Virtual Environments, 7(3), 225-240.

[31] Zahorik, P., & Jenison, R. (1998). Presence as Being-in-the-World. Presence: Teleoperators and Virtual Environments 7(1).

Tinker, Tailor, Software Engineer, Surgeon: Specialization in Software Systems Creation and Evolution

Stephen G. MacDonell
Auckland University of Technology
stephen.macdonell@aut.ac.nz

Diana Kirk
Auckland University of Technology
diana.kirk@aut.ac.nz

Abstract

This position paper advocates a change of mindset regarding how we perceive of and support those who develop and maintain software systems. We contend that a lack of explicit specialization is impeding our ability to deal effectively with the challenges that arise in the creation and evolution of software systems. Observations from the health sector lead us to reconsider the roles of the professionals involved.

Keywords: software engineering, software evolution, role specialization

1   Introduction

The burgeoning complexity faced by those building and maintaining software systems is widely acknowledged [1]. A consequence of this complexity is that, at all stages in a software life-cycle, an increasing range of expertise is required of the humans designing and managing the software product or service. The architect for a banking system with expertise in databases may be faced with designing an upgrade from a client-server to a web-based architecture. The need for improved security to address newly identified network vulnerabilities may require novice developers to rapidly learn and use protocol analysis techniques. An organization with an innovative software product may need to implement a 'light' approach to development with personnel that lack the understanding or ability to effectively support clients.

Why not simply implement training, initiate process improvement, hire new personnel, and so on - 'standard' responses to such demands? We contend that technologies are just too complex and fast-changing for training to keep up and the required expertise has become too broad for individuals to cope. If we consider the product or service alone, we must consider all of the *application area* (business, telephony, health...), the *product type* (web-based transaction, embedded real-time...) and the *operational environment* (desktop application, client server, distributed...). For a single organization, it is likely that the first (*application area*) would remain constant but equally likely that step changes in technology will force ongoing revisions of both *product type* and *operational environment*, with repercussions for both development and support personnel. If product and management processes are also considered in addition to the product or service, we must add the need to familiarize with a range of evolving techniques and tools, e.g. relating to requirements elicitation, change control, development, test and documentation. In addition to the techniques and tools that are 'standard' for the organization, project members must know how to deal with contextual issues. For example, an 'agile' group may have a process that assumes an on-site customer, and be left confused as to what to do on a project where there is no such access to customer.

There is a sense then that in software engineering (SE) there is more and more that we *need* to know. In broadening the body of knowledge, however, something has to give - we need more time to learn, to develop professionally, and/or we need to trade off breadth for depth. As a community we have not addressed this issue in an explicit and meaningful way. Basically, we have just tried to fit more in - into curricula, training courses, expectations for professional development. Such an approach is not sustainable, and we believe that the time has come to encourage greater disciplinary specialization. Elsewhere [3] we have suggested that the state of a software system at any point in time may be considered in terms of its health and well-being, just as we ourselves move through life stages characterized by states of wellness or illness. Taking this metaphor further can also provide insights into how we view the roles of those involved in the development and support of systems over time.

2 Systems and Specialization

Leveraging a human health metaphor can enable us to 'see' a software system in multiple ways, to depict a system depending on issues in focus. For instance, we may consider a software system to be young, mature, or aging [4]. Software systems in each grouping have particular attributes to consider when assessing their state of well-being. Similarly, we may view some systems as being ill and others as being well. Complementary to these approaches is a characterization of software systems according to their type. Systems that are hardware-embedded and operating in real-time are different to web-based transaction systems, and these again are different to personalized adaptive systems to be used on mobile devices. We can also consider systems in terms of their core domain or operating context - systems for business, systems for manufacturing, systems for assisted living.

We see such perspectives having direct analogues in the health sector, with what we believe to be useful implications. A person is also a complex system of systems that can be viewed in different ways - this may be in general classifications of age or maturity, or states of wellness/illness. We have infrastructural systems such as the cardiovascular, the musculo-skeletal, the respiratory, the nervous (somatic and autonomic) and the digestive. These could be seen as mapping to our network and database systems, to our computer and software architectures, to our application, interface and domain representations.

As humans we also comprise physiological and neurological systems that have physical and conceptual or logical elements - which may map to systems supporting workflow and enterprise activities.

All of these are valid perspectives for the consideration of software systems, each encourages and enables us to think about the system in a particular way, with a particular set of attributes in focus. However, in acknowledging that the various perspectives exist we are also acknowledging that they each represent part of the whole, and as such provide a potentially useful but inherently limited sense of the system. Furthermore, such systems and subsystems are not independent, they will interact in generally expected but sometimes unpredictable ways. Traditionally, software engineers have attempted to understand and cope with this breadth of perspective - an alternative is to allow for greater specialization with comparatively greater depth.

We take our model for role specialization from the health professions, an analogy also considered by Laplante [2] and others. Like software professionals, health professionals deal with a very complex entity in the human person. They are aware of interacting subsystems within that entity, as described above. In order to effectively understand, diagnose and treat that entity health professionals have adopted a specialized model. So, there are personnel with specific competencies in medicine or in surgery. Individuals may be experts in mental health, or provide specialist diagnostic support. Others take particular responsibility for assessment and rehabilitation. In the provision of services there is a clear distinction between primary and community care. In acknowledging that people of different ages may require different forms of support the health disciplines have gerontologists for those who are aging, and paediatricians for the young. In acknowledging the many subsystems that make up a person they have the nephrologist, the cardiologist, the gastroenterologist, the neurologist and the neurosurgeon. And there is finer granularity still when these perspectives intersect e.g. the paediatric nephrologist. This is not to say that generalists are not permitted. Quite the contrary, in fact, there are of course general practitioners in health. However, general practice is itself seen as a *specialty* - a specialty in breadth, requiring particular diagnostic skills, capabilities and ongoing professional development.

It may be suggested that we already have specialization in SE. This may be true to an extent, and certainly there are recognized specialist areas within the discipline, specialist degrees and courses that can be taken. In general, however, we continue to educate 'software engineers' - the equivalent to health's general practitioner. Our archival publication venues address the whole of the discipline, and job advertisements still seek software engineers albeit with particular skills and capabilities. However, these skills tend to reflect competence with the tools and technologies rather than with the application of tools to a particular class of (sub)system - like advertising for a surgeon with the requirement "Must have recent experience with Scalpel 3.0".

It may also be suggested that specialization is in fact a bad thing. For instance, specialization in SE has been criticized as causing an inability to "think holistically about the particulars of the problem that have been abstracted away, and that now may be the responsibility of no one in particular" [5]. We see this kind of specialization as a kind of 'functional specialization' that applies to the scoping of *activities*. For example, a developer may understand that (s)he must not modify design decisions, as this is the responsibility of the designer. We submit that the kind of specialization proposed in this paper is essentially different in that the scoping relates to the nature of the system, for example, to characteristics of the product or operating environment. Furthermore, the health model for team-based treatment helps to avoid inadequacies arising from specialized (and so limited) understanding. Multi-disciplinary teams (MDTs) work on a single entity in sequence and sometimes in parallel, often co-ordinated by a senior clinician. Members are called upon depending on the conditions encountered and the treatment plans chosen. Inter-disciplinary teams (IDTs) have an even greater degree of interaction - rather than a chain of individual specialists the group forms a network.

3 Implications

Carried through to software, the health model has particular implications for education and training. Prospective professionals would spend longer in formal education, but this would be increasingly practice-based, with internships the norm rather than the exception. A general foundation education would be followed by specialist learning and training, ongoing under compulsory development programs monitored by peer professionals. Some of this is done already, but it is not common practice. Professional development carries with it an expectation of self-reflection and learning in concert with a community-based approach to developing knowledge. For instance, the legal profession has case law, the medics have their grand rounds and evidence-based learning enabled by the Cochrane Collaboration. Of course, software is not a social service that governments support for the greater good, as they might for the health professions. But nor is law, yet we would not tolerate a law profession that did not learn as a community. Along with other changes in thinking and conduct [1, 2], disciplinary specialization may add to SE's growing professional standing.

References

[1] T. DeMarco. Software engineering: an idea whose time has come and gone? *IEEE Software*, Jul/Aug, 2009.

[2] P. A. Laplante. Professional licensing and the social transformation of software engineers. *IEEE Tech. Soc.*, Summer, 2005.

[3] S. G. MacDonell, D. Kirk, and L. McLeod. Raising healthy software systems. In *4th Intl Workshop Softw. Evoln Evolvability*, L'Aquila, 2008. IEEE CS Press.

[4] D. L. Parnas. Software aging. In *16th Intl Conf. Softw. Eng. (ICSE)*, Sorrento, 1994. IEEE CS Press.

[5] R. Schaefer. A rational theory of system-making systems. *Software Engineering Notes*, 31(2), 2006.

Geographic Vector Agents: Framework and Applications

Antoni Moore
University of Otago, New Zealand
tony.moore@otago.ac.nz

Kambiz Borna
Unitec Institute of Technology, New Zealand
kborna@unitec.ac.nz

Saeed Rahimi
University of Otago, New Zealand
saeed.rahimi@postgrad.otago.ac.nz

Abstract

There is a mismatch of most natural and anthropogenic geographic phenomena, which tend to have irregular geometry, and the regular-cell cellular automata often used as the basis for geo-simulation techniques. Geographic Vector Agents (GVA) were devised as a solution to this realism shortfall, containing agents with dynamic vector object geometric expression. This contribution introduces GVAs, defining its generic geometry, neighbourhood and state (attribute) classes. Three case studies that extend the GVA structure are outlined:

- a spatial agricultural land use model populated with geometric farm agents, farmer agents and a market agent
- an agent-based classification system for remotely-sensed image data in various contexts (spectral, elevation-based)
- moving object agents that are constrained in space and time (in a football game scenario) but subject to rules governing their behaviours.

Together, these examples demonstrate the flexibility and effectiveness of GVAs in modelling space-time phenomena.

1 Introduction

Benenson and Torrens [5, p. 4] have posed the question: "if modelled phenomena are an abstraction of real world phenomena, why should modelled objects differ from their counterparts in the real world?" This difference can refer to real-world phenomena that have imprecise or vague geometric boundaries (e.g. mountain) that can be difficult to incorporate in a digital spatial model. However, even if the real-world phenomena has precise boundaries, regular or irregular, dynamic spatial modelling featuring cellular automata (CA) and agents in particular have been slow to incorporate spatial units with a geometry to match reality.

CAs are usually square cell-based, with modelling over time generating global spatial order through application of rules to a local neighbourhood (typically a 3x3 window). Its simple spatial structure has underpinned complex models of urban phenomena in particular (e.g. [3]). However, the square cell-based division of space is limiting [5], with attempts to match the irregularity of real geographies including linear cells [47] and Voronoi polygons [42] as geometric advances, with increasing neighbourhood complexity attempted through Delaunay triangle links [41] and planar graphs [31]. VecGCA [29] is a land-use CA based on irregular vector objects corresponding to real-world entities, incorporating simulated change through rules applied to an irregular and dynamic neighbourhood.

Agents are computational objects with an embedded goal or set of goals, able to perform decision-making in order to progress towards those goals [27]. Geographically, agents have been applied at a meta-level to the integration of spatial environmental data [33]. Importantly, agent simulations can be explicitly spatial, with agents operating across geographic space. However, the prevalent CA-based agent simulation – often operating on a regular grid – had been geometrically naïve. This led to the introduction by Torrens and Benenson [45] of Geographic Automata Systems (GAS). One key aspect distinguishing GAS from CA, for one, is the means to store irregular geometry, with associated rules governing the change of geometry as

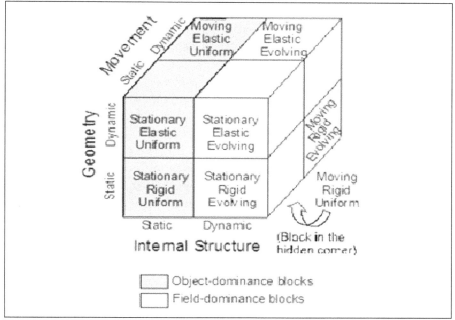

Figure 1: Types of dynamic geo-object derived from the three dimensions of geometry, movement and internal structure (Borna et al, 2014; derived from Goodchild et al, 2007).

a simulation progresses. Goodchild [17] have formalised this 'temporal variability in geo-objects' (Figure 1), implying eight types of geo-object arising from combinations of geometry, movement and internal structure (variation within the geo-object) properties. Of most relevance here are the fixed ('stationary'), non-fixed ('moving') and varying boundary ('elastic') elements.

Geographical Vector Agents (GVA) [20] is a generic spatial modelling framework developed to explore the scope of agents in this geometric sense. Although GVA and GAS have a shared approach in letting 'geography' drive the simulation, GVA aimed to address objects that change their geometric form over time, a group that although within the scope of GAS, was only briefly addressed: "...there are instances in which georeferencing is dynamic for the geographic automata that represent infrastructure objects, for example when land parcel objects are sub-divided during simulation." [45, p. 392]. Geometric change in GVA was achieved through application of object size and boundary complexity rules, with results compared favourably with selected real-world examples [20].

This chapter contribution will first define the GVA within the GAS framework, including a definition of a generic GVA class diagram. Three case studies documenting the evolution of GVA will then be presented: a theoretical agricultural land use model with spatial patterns emerging through the land function of polygonal farm agents; applying polygonal agents to remotely-sensed data to effect image classification based on spectral and elevation data, and; the application of point agents to movement analysis, from conceptual modelling, through to simulation of movement behaviours. Finally the contribution will be rounded off with a discussion and conclusion section.

2 Defining the GVA

The Geographical Vector Agent (GVA) is a type of Geographic Automata System (GAS; [45]), though physically and explicitly defined by its Euclidean geometry. It is a spatial agent able to modify its own geometry via rules, while using other rule sets to interact with other neighbourhood agents [20]. Defining GAS, there are seven elements:

$$G \sim (K; S, TS; L, ML; N, PN) \qquad (1)$$

- K distinguishes between spatially fixed (e.g. cell elements that do not change geometrically) and non-fixed objects (moving objects, objects with evolving boundaries). The GVA is non-fixed, its dynamics expressed through boundary manipulation or the ability to move as a whole (translation).

- L is the geometry (or "georeferencing convention") and ML are geometric rules. In GVA, this is an irregular (though could be regular) vector data structure, based on discrete coordinates and subject to projective transformation. This is the basis of representation of multifarious geographic phenomena, whether real-world objects or on a non-deterministic shape boundary. The agent has the ability to drive its own geometry, including location in space and boundary configuration.

- S is the set of states (or attributes) with TS being the transition rules effecting change of state. Accordingly, the GVA stores and manipulates its own spatial and aspatial attributes and their transition rules.

- R is the GAS spatial neighbourhood definition, with PN defining the rules that govern changes of neighbourhood. In GVA, this translates to a dynamic neighbourhood structure operating over a Delaunay triangulation of point representations of a group of agents (if two agents – represented by their centroid points – are linked by a triangle edge, then they could reasonably be neighbours).

It should be said that each rule set M_L, T_S and P_N is a function of all of geometry L, states S and neighbourhood N. The rules effect the change for the next time increment in the agent-based simulation, so the agent is as much a temporal agent as it is a spatial one.

3   Generic Model Definition and Implementation

3.1   GVA Definition

The UML diagram in Fig. 2 lays out the generic GVA class structure. There is one agent to which all other agents belong, VecContext, which is a container. Within this, the agent classes are structured in a hierarchy, with SimpleAgent as the top-level agent. SimpleAgent has two children, GeometricAgent and IndividualAgent. The Geometric-Agent class defines the types of agent that have a geometry, whether capable of changing that geometry or not. The IndividualAgent does not have a geometry in itself, but links to other agents of Geometric-Agent type that do.

The GeometricAgent class has two children, VecAgent and Maker-Agent. VecAgent is a dynamic geometry agent that can instantiate objects with properties and methods that can effect change on that geometry. The MakerAgent creates the VecAgents in the first place, having a simple geometry that does not change.

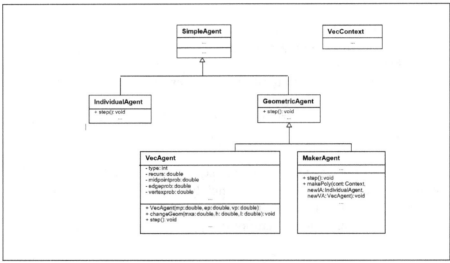

Figure 2: Simplified UML diagram of the GVA generic classes (adapted from Moore, 2011]

The VAGeometry and VAGeometryFactory classes (extensions of the Geometry and GeometryFactory classes from Java Topology Suite (JTS)) support the operations of the VecAgent (Fig.3). They include allocation definition in geographic space and methods such as mid-point / edge / vertex displacement that are instrumental in evolving a GVA. VAGeometry has three children, VAPoint, VALineString and

VAPolygon (extensions of the JTS classes Point, LineString and Polygon), which hold the geometric objects themselves. The VANeighbour class is used to access and make neighbourhood calculations on a linked network of VAGeometry objects / agents. The network has the object centroids as nodes and links constructed through a Delaunay triangulation of those nodes [41]. An object is a neighbour of another object if there is a link between their centroids. Extended neighbourhoods are also possible, say if an object is two or more links away on the network. This is achieved via a recursive function which calculates the neighbours of neighbours (and so on) of the focus object.

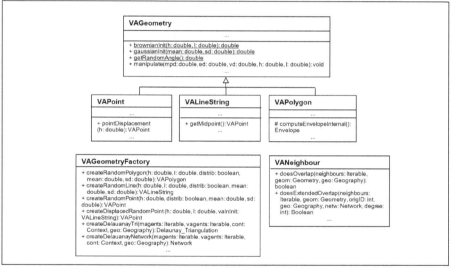

Figure 3: Simplified UML diagram of the GVA geometry and neighbourhood classes (adapted from Moore, 2011)

Most of the elements of the GAS definition are addressed by the GVA definition: geometry and geometry rules, neighbourhood (though neighbourhood rules are yet to be implemented); states and transition rules are introduced when this structure is extended towards a specific application (e.g. section 4).

3.2 GVA Implementation

The GVA was implemented in Java using the Repast (Recursive Porous Agent Simulation Toolkit) framework [26]. This development environment offered support for freeform vector data (i.e. had not necessarily pre-existed in a dataset) and the dynamic manipulation of its geometry, affording the evolution of non-deterministic GVA objects.

The "context" is the core data structure for Repast-S (Repast Simphony; [37]), represented by VAContext in Figure 2. The context can have geographic expression through a Geography "projection", a Euclidean space that can accommodate the geometry of GVAs. Other types of projection can exist in parallel with the Geography one (e.g. networks, grids). The generic GVA framework has a vector network projection that contains the neighbourhood links between agents, based on the Delaunay triangulation of agent centroids. Agents can simultaneously exist in the network and Geography projections. Furthermore, another network projection contains the links from IndividualAgents to the VecAgent or MakerAgent geometry currently associated with it (Association network). Figure 4 shows the relationship of the context and the three projections in the generic GVA model in Repast.

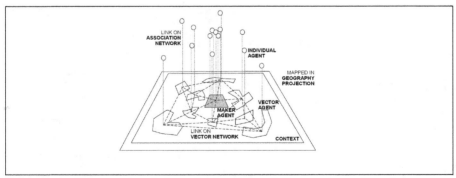

Figure 4: Schematic of the relationship of the context, the three projections and the agents that inhabit them, in the GVA generic implementation (adapted from [28]).

Finally, agents can use "watchers" to assess if the trigger for any given agent behaviour applies [26]. For example, watchers can assess the spatial environment to determine when conditions are optimal to initiate a GVA.

4 Agricultural land use case study

4.1 Background

An early case study for extending the GVA framework was that of an agricultural land use spatial model [28]. This is based on von Thünen's theory [46], [19, p. 8] that the agricultural land use around a central market would be a function of transportation cost and perishability of the land use type. This results in the following concentric ring zones ordered by distance to market (Figure 5).

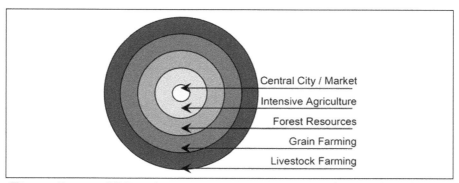

Figure 5: von Thünen's concentric ring pattern of agricultural land use (adapted from Moore, 2011).

Firstly, there is a ring of intensive agriculture (e.g. perishable fruits and vegetables, dairy), followed by forest resources, grain farming then livestock farming (a self-transporting resource, reducing transport costs). The assumptions of the theory are that there is a single market existing in an isolated state; the surrounding land is entirely flat and is of uniform fertility, and; there are no transport infrastructures, such as roads and rivers.

Since the 1950s, there have been numerous attempts to move von Thünen's theory from being a description to being an expression of the underlying process. Dunn (1954, cited in [23]) related production cost, market price and transportation cost in an equation to calculate land rent. The first computational model was created by Stevens [43], with a further model developed by Day and Tinney [15] and extended by Okabe and Kume [30]. The model that most informs this GVA implementation is the agent-based approach on a Cellular Automata grid implemented by Sasaki and Box [40].

The aim of the GVA simulation is to demonstrate the use and effect of irregular and dynamic geometry in this simple theoretical scenario. The hypothesis is that using start-up parameters adapted for the vector data model from Sasaki and Box [40], the end result will approximate to a concentric ring pattern around the central market or city, with the agricultural land use in the order of distance prescribed.

4.2 Definition and Implementation

Three sets of agents drive the von Thünen simulation (Fig. 6), extending the generic classes in Fig.2.

The FarmAgent extends the VecAgent class and therefore contains geometric rules (Fig 7) as methods. The first step is placing a point of random location in the model space as the "seed" for the farm. After turning into a line, then a triangle through the provision of two more points, a choice of three rules are applied to evolve the farm into a larger, more complex polygon. These rules implement the displacement of the midpoint of one of the polygon's edges, or the polygon edge itself, or one of the polygon vertices.

The FarmAgent has states (properties) pertaining to geometric manipulation, but most importantly, of its agricultural type (one of the types in Fig.5). Together, a collection of FarmAgents will form an agricultural land use pattern in model space, approximating the von Thünen concentric ring pattern (through local actions) if working correctly. The FarmAgent uses the Delaunay neighbourhood class to ensure that there is no geometric overlap with neighbouring farms.

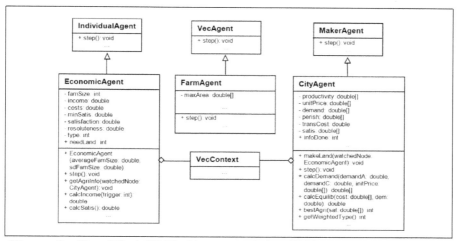

Figure 6: Simplified UML diagram of the GVA extensions required for the von Thünen simulation (adapted from [28])

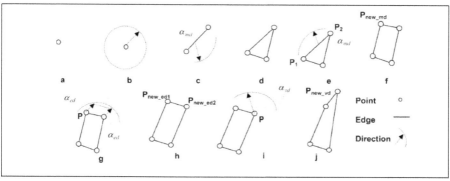

Figure 7: How a GVA shape is born and evolves in the agricultural simulation: (a - d) initialising a polygon via points and lines, (e – f) midpoint displacement, (g – h) edge displacement, (i – j) vertex displacement (from [28]).

The CityAgent extends the MakerAgent class and has an underlying geometric footprint. Its main purpose is to function as the market for the model, a role that includes (through states and associated methods) the initialisation, storage, and updating of each agricultural type's productivity, price, demand, perishability, transport cost and farmer satisfaction. The CityAgent broadcasts this information to every EconomicAgent (the third group of agents, described next) and allocates those agents to farms (creating FarmAgents if necessary) or the city, according to their desire.

The EconomicAgent represents the human decision maker linked to either the CityAgent or a FarmAgent via the Association network. Through states and methods, it initialises, stores and updates its own income, cost, satisfaction, resoluteness and city / agricultural type.

An additional Repast network projection economically links each EconomicAgent to the CityAgent, the EconomicAgent receiving up-to-date market information, acting accordingly as a result. For example, if a city dweller calculates that they will be better off farming, then they can signal the CityAgent to arrange this for them. Alternatively, a switch from a farm to the city or an alternative agricultural type can be facilitated. The process is summarised in Figure 8.

4.3 Outcomes

The start-up parameters for the simulation were adapted from those for Sasaki and Box's [40] cellular automata model, with the exception of geometric parameters, unique to GVA. Several model runs with varying parameter values were made to find an optimal level between the simulation being too dynamic or too static, whilst generating results that approximated most to the concentric ring pattern in Fig. 5. Two realisations are shown in Fig. 9. As anticipated, the concentric ring pattern is not reproduced exactly, but if farms are ordered by average distance to market, grouped by agricultural type, there is at least the approximately correct sequence (from intensive agriculture to livestock farming, though forestry and grain farming are in reverse order) emerging.

The agricultural land use scenario demonstrates a key principle of spatial agents, that a plausible global pattern can be generated through localised interaction. However, there are inefficiencies in the implemented model (e.g. an agent polygon will always strive to expand into empty space, even if it is very limited), neighbourhood and

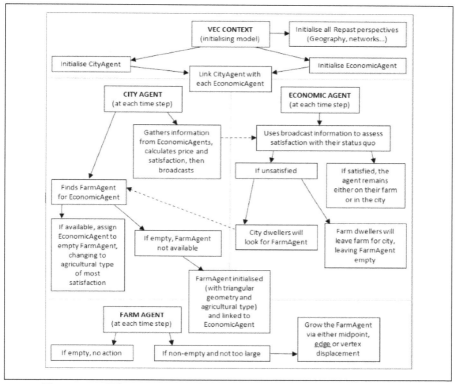

Figure 8: The process of the GVA von Thünen simulation

associated rules are underutilised, and there is a need to move from this theoretical agricultural scenario into a contemporary real-world one.

5 Image Classification Case Study

5.1 Background

A later case study used GVA to classify and extract geographic objects (geo-objects) from remotely sensed images, in contrast to conventional classification methods that use pixels or image objects [9, 10]. Image-

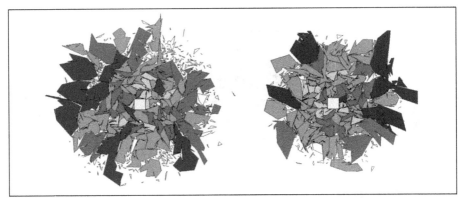

Figure 9: Two realisations of von Thünen GVA output (refer to Figure 5 for comparison with theory and legend; white = empty farms)

objects are a group of connected pixels having a similar digital number (DN), enabling information beyond the spectral to be factored into the classification task, such as shape, texture, edge [6, 22, 44] and the spatial and thematic knowledge of geo-objects [21]. This can increase the accuracy of classification results, especially with a high spatial resolution (H-resolution) image and when compared to the limited information enabled with classification of the pixel alone.

This process is the basis of Geographic Object-Based Image Analysis (GEOBIA) [6, 21, 25], which typically uses a linear two-stage procedure [6]. The first step is image segmentation, merging individual pixels to create image objects. In the second step, image objects are analysed and classified into a set of classes. However, although GEOBIA is consistently and accurately effective, image objects suffer from subjectivity due to scale [22, 16], are not consistently homogeneous [44, 16] and might not represent the final objects of interest – exacerbated by no direct connection to the corresponding geo-objects [2].

A number of presented solutions have tried to address these limitations by applying expert knowledge during image classification. Baatz et al. (2008) introduced geo-object-informed rules to label segmented image objects (e.g. tree areas are designated by Digital Surface Model

– DSM – values more than a specific threshold). Blaschke et al. [7] added a multi-scale segmentation process to this intelligent linking of image objects and geo-objects. Hofmann et al. [25]) introduced agents to the process, so that after the initial segmentation and classification steps, the generated Image Object Agents (IOA) can re-segment themselves or merge with a neighbouring IOA during the classification process.

Although these approaches allow image objects to change their geometry and use expert knowledge during the classification step, geometric changes only include re-segmenting or merging image objects. In other words, these methods ignore the local geometric changes between image-objects. Fundamentally, if geo-objects are key to the GEOBIA approaches [7, 13], then why is there a bigger focus on image objects, rather than on geo-objects? This echoes the question [5, p. 4] that opens this chapter. In the following section, we will analyse and discuss the different elements of GVA that address geo-objects in images.

5.2   Definition and Implementation

The elements of the GVA for image classification are summarised and displayed in Figure 10. The image raster itself has a spatial extent and n spectral bands (layers). A feature space of n-dimensions is created, defined and filled by the DNs of pixels in each band. A vector space, which is a two-dimensional space of the spatial extent of the image, is initialised, and will accommodate the geometry of agent-derived geo-objects. The agent's sensor takes in spectral, spatial and temporal information from the feature and vector space, analysing and classifying pixels in the feature space. Geo-objects are dynamic, determining their own actions in relation to their environment and other geo-objects.

The geometry of this GVA implementation is based on polygon units comprising vertices that define the polygon boundary $[\delta X]_{G}VA$. The geometry is subject to a set of decision rules which can be divided into two groups. Internal rules are defined by the characteristics of the GVA, allowing a GVA to dynamically change its own geometry in

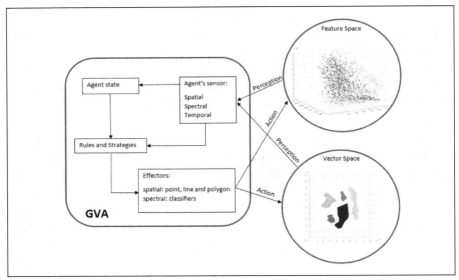

Figure 10: Schematic diagram of the model architecture of the image classification GVA including state, sensor, rules and strategies, and effectors (adapted from [11])

a process of iterative evolution (Figure 11).

External rules involve the interaction with the geometry of other GVAs and can lead to the birth, death as well as geometric change of interacting GVAs. For example, a GVA can remove and transfer a vertex from other GVAs (Figure 12a-c illustrates how a GVA labelled as shadow can split a GVA labelled as road into two).

The state or class (label attributes such as road and shadow in the previous example) of each GVA is determined by the class, geometry of the geo-object, and the list of neighbour geo-objects. It is represented by X_{GVA}, a connected subset of labels stored with the pixels belonging to the GVA, as well as the spectral signature of each class, contextual information or structural descriptors. There are transition rules to find, evaluate and update classes and attributes, such as the statistical rule Maximum Likelihood (ML) or Support Vector Machine (SVM), applied to the feature space to identify the class of each pixel,

122

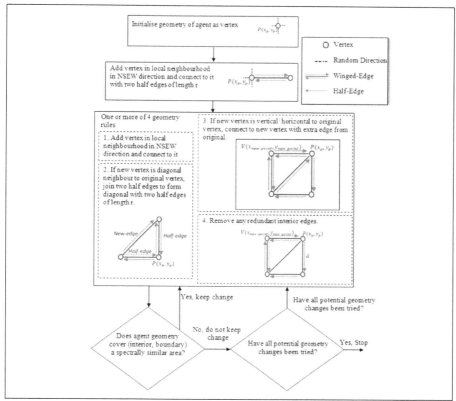

Figure 11: The creation of a geo-object and changes through time in the vector space. Note that each vertex corresponds to the centre of a raster cell in the remotely sensed image being classified and is thus part of a regular lattice (from [9]).

updating the corresponding GVA class.

The neighbour geo-objects are explicitly determined based on the Euclidean distance separation with the GVA in the vector space. To update the neighbour geo-object list, GVAs apply neighbourhood rules to find the GVAs that are in the local neighbourhood. Neighbours can be contiguous or non-contiguous, the latter subject to extended neighbourhood rules. Figure 12d-e demonstrates a contiguous exam-

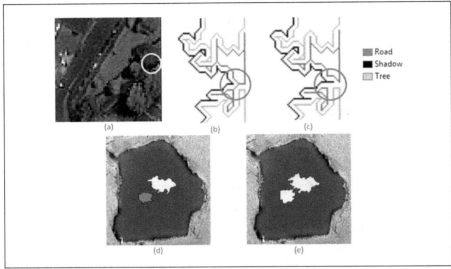

Figure 12: Examples of GVA rules in action (a-c) In this Dunedin IKONOS image extract, a relatively dark pixel was initially found to match the road GVA, but subsequently changes to shadow, leading to the birth of a new road GVA (from Borna et al. 2016); (d-e) Another IKONOS extract of a water body has elicited evolution of one pond GVA (red) and one lake GVA (yellow), which become contiguous neighbours and merge to form a single lake GVA.

ple of a merging / killing process as a pond GVA and lake GVA unite to create a single lake GVA, also invoking a threshold size rule for the definition of lake.

Operating over this geometric, attribute and neighbourhood structure, these GVAs follow a sequence to identify geo-objects from the image: growth, development, construction and production. Implementation-wise, the content of the image classification GVA was developed in Repast, extending the generic GVA classes.

5.3 Outcomes

GVA-based image classification was tested with a rural Otago World-View-3 image subset (140x90 pixels of 8 spectral bands – red, red edge, coastal, blue, green, yellow, near-IR1 and near-IR2 – at 1.2m resolution) chosen for land cover complexity (bare soil, grass, lake, pond, river and shadow; Figure 13a). To train the SVM classifier used to implement the transition rules, nine labelled pixels were used for each cluster (results of classification in Figure 13b). The results of the first (growth) and last (production) GVA steps are shown in Figure 13c-d.

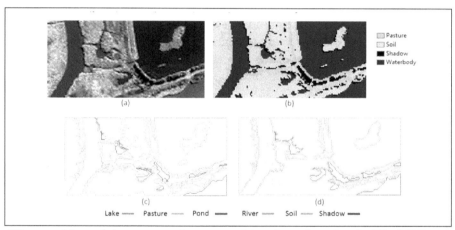

Figure 13: (a) true colour combination band of WorldView-3 multispectral imagery (Digital Globe Foundation, www.digitalglobefoundation.org); (b) SVM classified image based on initial label samples; (c) GVA-extracted geo-objects after the initial growth step, and (d) GVA results in the final production step.

In this case, external rules on size and elongation of water bodies was used to distinguish between lake, river and pond geo-objects. The GVA model can therefore accurately identify and extract geo-objects which have similar DNs.

The obtained results indicate the ability of the GVA model to provide a unified and dynamic geometry for geo-objects during an image classification process. This structure allows the geo-objects to be directly extracted from an image without using image-objects, such as with previous classifiers. In this way, the emphasis shifts to geo-objects, which can be detected and modelled in the same way a human interpreter manually identifies and extracts geo-objects from an image.

6 Movement Analysis Case Study

6.1 Background

The third case study is included to indicate where our research on vector agents is currently at, turning this technology to the patterns and causality of movement, as described by multitemporal individual point data, and reaching beyond the ABM-based account featured in this chapter up to now [34, 35]. The current abundance and availability of movement data has encouraged researchers to prioritize finding credible narratives that account for observed movement behaviours. This, together with a widely-accepted model would potentially enable both theoretical and applied Computational Movement Analysis (CMA) to shift their objectives from describing 'what happened' towards explaining 'how and why it happens' (i.e. the process and causes)

In practice, Graphical Causal Modelling (GCM) and Agent-Based Modelling (ABM) both claim to explore scientific causally-relevant evidence. Whilst both methodologies are increasingly being developed and used in different fields, their utility for scientific understanding is the subject of many recent theoretical and practical dialogues. ABMs are accused of being too simplistic or, at best, black-boxes that run under untestable assumptions. GCMs are criticised for being highly dependent on data, and incapable of dealing with complex mechanisms. They also both rely on theoretical, substantive, and domain-specific knowledge to show that the assumptions on which the causal identification strategy relies are satisfied (see [12]) on the critiques against these two approaches).

A potential solution integrates both approaches into data-driven agent architectures, where the modellers' initial assumptions, agents' behaviours, and the outcomes could be constructed and validated based on collected observations [4, 8, 14, 24]. In an attempt towards such a unifying concept, Rahimi et al. [34] suggested a three-level cognitive framework (Figure 14) for explaining movement mechanisms along the association, intervention, and counterfactual levels of operation (see [32]) for description of these levels). This is founded on a conceptual agent-based representation of the real world (see [34] for a detailed discussion) in which the movement mechanisms could be represented through reconstructing the causal effects of moving entities' indigenous attributes, environmental actors, and other autonomous moving agents' actions. These three conceptually different zero-, first-, and second-order causal factors are observed, analysed, and validated over space, time, or both space and time.

The first stage in the implementation of this conceptual cube covered the 'Association' level of operation (i.e. the front row of the cube) and an example scenario that was tightly constrained in both time and space. Professional sport analysts are increasingly showing interest in looking beyond match statistics and average team performances to model movements of individuals [18]. Among them, the motion of football players is perhaps one of the most intensively observed and thoroughly discussed processes [39, 36]. Modelling the complex decision-making mechanism underlying movement of football players would be a relevant research endeavour and a way to start demonstrating the conceptual model.

6.2 Definition and Implementation

A vector agent-based movement simulation of a football game has been developed in NetLogo, based on the assumption that movement decisions are caused by three sets of factors (Figure 14):

- 'Zero-order' factors, characterising the players' endogenous capabilities. These are 'Stamina,' 'Energy,' 'Pace,' 'Agility,' and 'Shooting' abilities (or attributes).

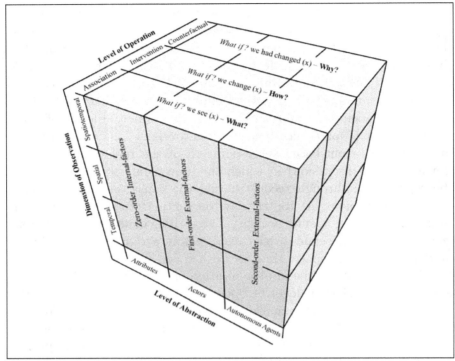

Figure 14: The conceptual representation of enquiries that an intelligent point vector-agent could possibly ask and answer by considering its: state of attributes, interactions with the environmental actors, and relations with other autonomous agents over space and time [34]

- 'First-order' causes, representing the environmental actors. These actors include an imagined hard-bounded-box around each player's role-area, the elements that shape the football pitch, and the Goals. The ball is also considered as an environmental actor, as it does not move due to an autonomous-made decision.

- 'Second-order' causal factors, indicating the interactions among autonomous moving agents (players).

Figure 15 shows the player autonomous agents, as well as their respective hard-boundary role-areas, pitch, goals and ball (environmental actors).

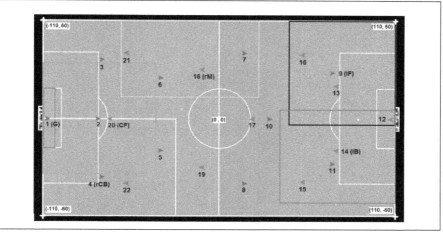

Figure 15: Autonomous agents (players) and environmental actors (player role-boxes, pitch, goals, ball) in the football simulation [35]).

Figure 16 depicts the rule structure that affects player decision-making. An autonomous movement decision involves selecting a direction and speed (can be zero in case of stopping). Autonomous agents choose these two parameters with regard to their objectives of movement (actions). To select an action, players try to find out 'who possesses the ball' first and foremost. If it is not them or one of their teammates, they find a location at which they can possibly intercept and possess the ball. This is a function of their current abilities and the current velocity of the ball. If this location is inside their allocated role-area, they attempt to get the possession of the ball. Otherwise, they decide to watch (closely move with) the nearest opponent.

When players themselves possess the ball, they choose either to carry or shoot the ball towards the opponent's goal, or to pass it forward to one of their teammates. To do so, they consider their distance to opponent players, to the goals, and to the centre of their

129

own role-box alongside with their current abilities. But if one of their teammates possesses the ball, players try to open-up the space by moving forward and towards one of the side lines of the pitch.

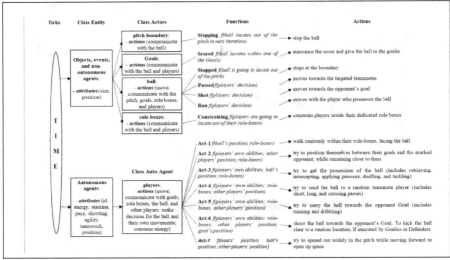

Figure 16: The structure of the model ([35]; adapted from [1])

Lastly, to incorporate a degree of freedom to the game, players have been designed to move randomly around the centre of their role-area, when their distance to the ball is more than a certain length. One remaining unknown procedure is how they incorporate their capabilities in their decisions. According to the model configuration, at each time interval, players perceive their environment and consider their pace, agility, and shooting abilities together with their current level of energy (reduced over time as a function of stamina) to determine what to do and where to go. This configuration together with the dynamic mechanisms of the model explicate the three types of interaction: among auto-agents (e.g., a player and teammates), between auto-agents and actors (e.g., players and the ball), among actors (e.g., the ball and pitch's boundaries).

6.3 Outcomes

In the light of the variety and complexity of model outcomes, subject to individual, role-based, team, and aggregated performances over space and time, three representative types of outcome were chosen. These were: (i) ball possession patterns, (ii) the frequency of executed actions (objectives of movement: Act-1 to Act-7), and (iii) players' movement relative to the ball and the closest opponent player. Five configurations of model parameters were used to generate outcomes, varying energy consumption (E), formation (F), and marking strategy (M):

- E – consumed, F – 3-5-2, M – man-to-man (C1);

- E – not consumed, F – 3-5-2, M – man-to-man (C2);

- E – consumed, F – 3-5-2 (Team A) 4-4-3 (Team B) , M – man-to-man (C3);

- E – consumed, F – 4-4-3 (Team A) 3-5-2 (Team B), M – man-to-man (C4);

- E – consumed, F – 3-5-2, M – zonal marking plan (C5).

For a single model run, Figure 17a shows intra- and inter-team ball possession patterns for most configurations. An example role-based pattern emerging is Team A's strong link between midfielders and strikers and Team B's emphasis on the defender-midfielder link, seemingly due to individual attributes (e.g. Team B's strikers are more self-sufficient due to their high average stamina). Figure 17b-c shows the frequency of selected executed actions over all configurations for each player (refer to Figure 16 for the labelling of Act-1 to 7; player numbers are in order of roles – goalkeeper, defender, midfielder, striker – for both teams), demonstrating a consistency from player to equivalent player in the opposing team (roles demonstrate consistent groupings, too).

Figure 17d shows the movement and distance relative to the ball for one player only (Player 10, the most active agent). One obvious

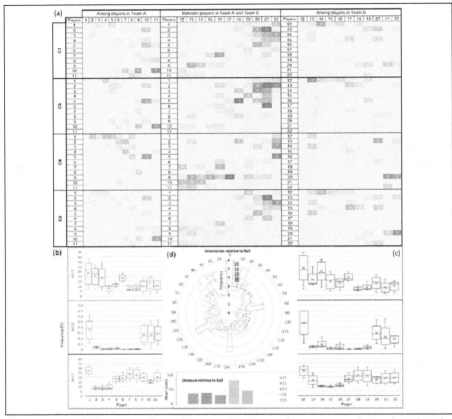

Figure 17: Selection of result visualisations for team, role and individual aspects: (a) Changing ball possession among players and between teams for selected configurations. The green colour indicates successful passes among teammates. The red cells show the accumulated number of successful attempts to get the ball (retrieves, tackles, interceptions etc.) from the opposition; (b + c) The variation of selected executed actions (attempting ball possession, shooting, running into space) over all configurations, at individual-level. The lower values in Act-6 (the shooting case) is due to the requirement for possession of the ball, making it relatively rare; (d) Relative distance and orientation towards the ball for Player 10 only. Some values for the dominant 'north' orientation (i.e. towards the ball) beyond the plot extent are represented as numbers according to colour-coded configuration.

trend that emerges is that the closer the player is to the ball, the more consistently oriented towards the ball that player is. A similar trend can be observed with Player 10 and its closest player (data not shown) but with orientation less intensively focussed, even though average distance to the other player is less than average distance to the ball.

Overall, the variation of team formation has a bigger impact on the emergent outcome than the energy consumption rate and marking strategy, possibly due to the high influence of role-boxes on the running of the simulation, especially for the defender role. Ultimately, the movement behaviour of players does not seem to change radically across all the model settings, which underpins the robustness of the simulation. The next stage of development is to expand the agent implementation from the 'What?' (association) level of operation in the conceptual model (Figure 14) into the 'How?' (intervention) and 'Why?' (counterfactual) levels.

7 Overall Discussion And Conclusion

The GVA has been designed and implemented as an attempt to address the challenge of irregular geometry often possessed by real-world natural and anthropogenic phenomena. The agent has dynamic vector object geometric expression, which has been successfully applied to scenarios of differing spatial scale, temporal scale, attributes and neighbourhoods, as well as rules relating to all of these.

The agricultural land use scenario had agents deployed that operated at a regional (city market and surrounding farms) spatial scale and a temporal scale in the order of years to decades. A small number of states were possible (intensive, forest, grain, livestock, empty) and the neighbourhood had a fixed definition (up to two Delaunay triangle links, which was nevertheless adaptive to the density of agents). The agent rules underpinning the simulation were relatively simple geometrically but transition rules governing change of state were more sophisticated, based on many attributes.

The image classification case study operated directly on remotely-sensed imagery to derive geo-objects of a fine spatial scale, though it is possible to successfully run the model on data of coarser scales. Temporally, the agents were applied to an image of a single fixed time. The number of states was adaptive, therefore flexible, to the number of classes sought in the image classification, and neighbourhood was defined by pixel adjacency (from information stored as a planar graph). The geometric rules were similar to the agricultural case study, but more complex, governed by all the unique ways in which the agent can change its boundary relative to the underlying pixel grid. Transition rules were relatively straightforward, based on available spatial data and enacted through conventional classifiers e.g. SVM or ML. However, with this case study, there was neighbourhood rule development, with planar graph representation defining the boundaries of classifier agents and those of neighbouring agents (often a function of the agent state e.g. of trees and their shadow, two agents that work in tandem).

The movement case study was the first stage of the implementation of a conceptual model operating along Level of Abstraction (self-attribute to autonomous agent), Dimension of Observation (space, time, and space-time) and Level of Operation ("what", "how", and "why") axes. The football case study was run at large (pitch) spatial scale and short (length of a game) temporal scale. There was a fixed amount of states for the player agent (depending on whether the agent possessed the ball or not, and how close the player was to the ball, leading to seven possible actions for the agent. Neighbourhood is defined by proximity of the player to other players (own and opposing team), and the ball. Geometric rules pertained to movement of point agents only, linked to actions from random walk to more directed movement relative to other players, the ball and the goals. Transition rules were rather more complex, often combining with the geometric rules to perform the typical actions in a football game (blocking, tackling, dribbling, passing, shooting). The agent player neighbourhoods (by distance) were fixed (i.e. no rules).

Seen as a GAS, there are challenges that remain for GVAs. As well as the large variety of applications to target (one of the more intriguing applications is having agents as intelligent guides through visual analytics tools of Big Data – [38]), there are tasks such as multi-scale modelling as well as implementations based on lines and true 3D, geometrically and temporally. The role of neighbourhood has not been explored extensively up to now. Possible extensions here would explore rules that changes neighbourhood, as dictated by the state of the model run (e.g. adaptively modifying the degree of Delaunay neighbourhood in the agricultural land use scenario). Even the existing neighbourhood implementations could be improved: in the image classification case study, creating a new vertex for a polygonal vector agent involves checking the four cardinal directions for each existing vertex of the agent, a procedure that could be streamlined.

Finally, for the state or attribute-based functions of GVAs, the immediate way forward is heralded by the movement case study, in modelling causality by moving from the "What" functionality of agents to the "How" and the "Why" [35]. Such modelling towards counterfactual scenarios would be enabled by more cognitively complex agents, equipped with graphical models (e.g. graph-based models that use Bayes or Dempster-Shafer theorems).

Acknowledgements

This chapter is dedicated to Martin Purvis, Professor of Information Science with fingers in an impressive amount of pies, including spatial agents. Thanks for all your help, guidance and wisdom, especially back in the day when I was starting out as an InfoSci lecturer.

Thanks also to Yasser Hammam, Peter Whigham and Pascal Sirguey for helping with the development and testing of Geographical Vector Agents in various forms and for various purposes.

References

[1] Ahearn, S.C., Smith, J.L.D., Joshi, A.R., Ding, J. (2001). TIGMOD: an individual-based spatially explicit model for simulating tiger/human interaction in multiple use forests. Ecological Modelling, 140, 81–97.

[2] Baatz, M., Hoffmann, C., Willhauck, G. (2008). Progressing from object-based to object-oriented image analysis. In: T Blaschke, S Land, G J Hay (eds.). Object-based Image Analysis: Spatial Concepts for Knowledge-driven Remote Sensing. Springer, Berlin. 29–42.

[3] Batty, M. (2001). Cellular dynamics: modelling urban growth as a spatial epidemic. In: M. M. Fischer and Y. Leung (eds.). GeoComputational Modelling: Techniques and Applications. Advances in Spatial Science. Springer. 109-141.

[4] Bell, D., Mgbemena, C. (2018). Data-driven agent-based exploration of customer behavior. Simulation, 94, 3, 195–212.

[5] Benenson, I., Torrens, P. (2004). Geosimulation: Automata-Based Modelling of Urban Phenomena. Wiley, England.

[6] Benz, U.C., Hofmann, P., Willhauck, G., Lingenfelder, I. & Heynen, M. (2004). Multi-resolution, object-oriented fuzzy analysis of remote sensing data for GIS-ready information. ISPRS Journal of Photogrammetry and Remote Sensing, 58, 3, 239-258.

[7] Blaschke, T., Hay, G.J., Kelly, M., Lang, S., Hofmann, P., Addink, E., Feitosa, R. Q., van der Meer, F., van der Werff, H., van Coillie, F. & Tiede, D. (2014). Geographic object-based image analysis–towards a new paradigm. ISPRS Journal of Photogrammetry and Remote Sensing, 87, 180-191.

[8] Boero, R., Bravo, G., Castellani, M., Squazzoni, F. (2010). Why Bother with What Others Tell You? An Experimental Data-Driven Agent-Based Model. Journal of Artificial Societies and Social Simulation. 13, 3, 6. doi:10.18564/jasss.1620.

[9] Borna, K., Moore, A., Sirguey, P. (2014). Towards a vector agent modelling approach for remote sensing image classification. Journal of Spatial Science, 59, 2, 283-296.

[10] Borna, K., Moore, A. B., Sirguey, P. (2016). An Intelligent Geospatial Processing Unit for Image Classification Based on Geographic Vector Agents (GVAs). Transactions in GIS, 20, 368-381.

[11] Borna, K. (2017). Geographic Vector Agents from Pixels to Intelligent Processing Units (Thesis, Doctor of Philosophy). University of Otago.

Retrieved from http://hdl.handle.net/10523/7183

[12] Casini L., Manzo, G. (2016). Agent-based Models and Causality: A Methodological Appraisal. Linköping University, Sweden.

[13] Castilla, G. & Hay, G. (2008). Image objects and geographic objects. In: T. Blaschke, S. Land and G.J. Hay (eds.) Object-based Image Analysis: Spatial Concepts for Knowledge-driven Remote Sensing. Springer, Berlin. 91-110.

[14] Chen, S.-H. & Venkatachalam, R. (2017). Agent-based modelling as a foundation for big data. Journal of Economic Methodology, 24, 4, 362–383.

[15] Day, R.H. & Tinney, E.H. (1969). A dynamic von Thünen model. Geographical Analysis, 1, 137-151.

[16] Gao, J. (2008). Digital analysis of remotely sensed imagery. McGraw-Hill Professional.

[17] Goodchild, M.F., Yuan, M & Cova, T.J. (2007). Towards a general theory of geographic representation in GIS. International Journal of Geographical Information Science, 21, 3, 239–260.

[18] Gudmundsson J. & Horton M. (2017). Spatio-temporal analysis of team sports. ACM Comput. Surv. 50, 1–34.

[19] Hall, P. 1966. Isolated State: an English edition of 'der Isolierte Staat'. Pergamon Press.

[20] Hammam, Y., Moore, A. & Whigham, P. (2007). The dynamic geometry of geographical vector agents. Computers, Environment and Urban Systems. 31, 5, 502-519.

[21] Hay, G. J. & Castilla, G. (2008). Geographic Object-Based Image Analysis (GEOBIA): A new name for a new discipline. In: T. Blaschke, S. Land and G.J. Hay (eds.) Object-based Image Analysis: Spatial Concepts for Knowledge-driven Remote Sensing. Springer, Berlin. 75-89.

[22] Hay, G. J., Castilla, G., Wulder, M. A. & Ruiz, J. R. (2005). An automated object-based approach for the multiscale image segmentation of forest scenes. International Journal of Applied Earth Observation and Geoinformation. 7, 4, 339-359.

[23] Henshall, J. (1967). Models of agricultural activity. In: R. J. Chorley and P. Haggett (Eds.) Socio-economic models in geography, Methuen. 425-458.

[24] Herd, B.C. & Miles, S. (2019). Detecting causal relationships in simulation models using intervention-based counterfactual analysis. ACM Trans. Intell. Syst. Technol. 10. https://doi.org/10.1145/3322123

[25] Hofmann, P., Lettmayer, P., Blaschke, T., Belgiu, M., Wegenkittl, S., Graf, R., Lampoltshammer, T. J. & Andrejchenko, V. (2015). Towards a framework for agent-based image analysis of remote-sensing data. International Journal of Image and Data Fusion, 6, 2, 115-137.

[26] Howe, T., Collier, N., North, M., Parker, M. & Vos, J. (2006). Containing agents: Contexts, projections and agents, In: Proceedings of the Agent 2006 Conference on Social Agents: Results and Prospects. Argonne National Laboratory.

[27] Luck, M., D'Inverno, M. & Munroe, S. (2003). Autonomy: Variable and generative. In: H. Hexmoor, C. Castelfranchi, and R. Falcone (Eds.) Agent Autonomy. Kluwer Academic Publishers. 9-22.

[28] Moore, A. (2011). Geographical Vector Agent-Based Simulation for Agricultural Land-Use Modelling In: D.J. Marceau and I. Benenson (Eds.) Advanced Geosimulation Models. Bentham Science, United Arab Emirates. 30–48.

[29] Moreno, N., Wang, F. & Marceau, D. (2009). Implementation of a dynamic neighbourhood in a land-use vector-based cellular automata model. Computers Environment and Urban Systems, 33, 1, 44-54.

[30] Okabe, A & Kume, Y. (1983). A Dynamic von Thünen model with a demand function. Journal of Urban Economics, 14, 3, 355-369.

[31] O'Sullivan, D. (2001). Graph-cellular automata: A generalised discrete urban and regional model. Environment and Planning B, 28, 687-707.

[32] Pearl, J. & Mackenzie, D. (2018). The book of why: The new science of cause and effect. Basic Books.

[33] Purvis, M., Cranefield, S., Ward, R., Nowostawski, M., Carter, D. & Bush, G. (2003). A multi-agent system for the integration of distributed environmental information. Environmental Modelling and Software, 18, 565-572.

[34] Rahimi, S., Moore A.B. & Whigham, P.A. (2020a). Beyond objects in space-time: Towards a movement analysis framework with 'How' and 'Why' elements. Manuscr Submitt Publ

[35] Rahimi, S., Moore A.B. & Whigham, P.A. (2020b). A Vector Agent approach to movement modelling and analysis. Manuscr Submitt Publ

[36] Rein, R. & Memmert, D. (2016). Big data and tactical analysis in elite soccer: future challenges and opportunities for sports science. Springerplus 5, 1–13. https://doi.org/10.1186/s40064-016-3108-2

[37] Repast. (2016). Recursive Porous Agent Simulation Toolkit. [Online] Available: https://sourceforge.net/projects/repast/ [Accessed Nov.

27th, 2020].

[38] Robinson, A. C., Demšar, U., Moore, A. B., Buckley, A., Jiang, B., Field, K., ... & Sluter, C. R. (2017). Geospatial big data and cartography: research challenges and opportunities for making maps that matter. International Journal of Cartography, 3(sup1), 32-60.

[39] Sarmento, H., Marcelino, R., Anguera, M.T., Campanico, J., Matos, N. & Leitao, J.C. (2014). Match analysis in football: a systematic review. J Sports Sci., 32, 20, 1831–1843. https://doi.org/10.1080/02640414.2014.898852

[40] Sasaki, Y. & Box, P. (2003). Agent-based verification of von Thünen's location theory, Journal of Artificial Societies and Social Simulation, 6, 2, 9. [Online] Available: jasss.soc.surrey.ac.uk/6/2/9.html [Accessed Nov. 27th 2020].

[41] Semboloni, F. (2000). The growth of an urban cluster into a dynamic self-modifying spatial pattern. Environment and Planning B, 27, 4, 549-564.

[42] Shi, W. & Pang, M.Y.C. (2000). Development of Voronoi-based cellular automata: An integrated dynamic model for geographical information systems', International Journal of Geographical Information Science, 14, 5, 455-474.

[43] Stevens, B. (1968). Location theory and programming models: the von Thünen case. Papers of the Regional Science Association, 21, 19-34.

[44] Tian, J. & Chen, D.-M. (2007). Optimization in multi-scale segmentation of high-resolution satellite images for artificial feature recognition. International Journal of Remote Sensing, 28, 20, 4625-4644.

[45] Torrens, P. & Benenson, I. (2005). Geographic Automata Systems. International Journal of Geographic Information Science, 10, 4, 385-412.

[46] von Thünen, Johann Heinrich. (1826) *Der isolierte Staat*

[47] Wahle, J., Neubert, L., Esser, J., & Schreckenberg, M. (2001). A cellular automaton traffic flow model for online simulation of traffic. Parallel Computing, 27, 719-735.

SELF-SOVEREIGN DIGITAL INSTITUTIONS

MARIUSZ NOWOSTAWSKI
Norwegian University of Science and Technology
did:ion:EiAeJa_4kL_EGfUk1-hSHFworj4S9OFf3zv8uAOdj3AUtw

ABYLAY SATYBALDY
Norwegian University of Science and Technology
did:ion:EiBCdOhplxLsWBN3muYxAXPoKoMMaFM6NZtpxydDnV8AGA

Abstract

We evaluate the state of art in the contemporary open source socio-technical systems with the focus on the decentralised finance (DeFi), present the fundamental assumptions, drivers, achievements as well as pitfalls of the current experiments. We note that the contemporary blockchain-based approaches that rely on proof of work or proof of stake and the pseudo-anonymity of the participating actors on all layers of the value stack fall short in delivering on the promises related to privacy, openness, governance, interoperability and decentralisation. To re-frame and re-think the fundamental assumptions, we focus on the open innovation models and re-use based on sound and well-formulated principles, open systems interconnect (OSI) layers and metaphors, and open, royalty-free standards. We sketch the notion of a self-sovereign digital institution (SDI), and argue, that for digital institutions to be self-sovereign and to be able to mimic the level of social organisation of existing contemporary institutions in the physical world, the combination of decentralised governance including decentralised sanctions and law enforcement, decentralised marketplace and decentralised,

---

Note, DID identifiers can be resolved through universal DID resolver **https://uniresolver.io**. For the ION-based DIDs, one can use ION explorer **https://identity.foundation/ion/explorer/**

user-centric digital identity management needs to be combined together. None of this components in isolation is sufficient in achieving the outlined goals. The paper introduces the necessary building blocks for Self-sovereign Digital Institutions: the notion of the institution, including rules and sanctions, the notion of decentralised market and finance, decentralised identity and self-sovereign digital governance. We use Lessig's four dimensions: Code, Law, Norms and Markets, as the fundamental elements to be considered when discussing open decentralised systems in general, and decentralised financial systems and institutions in particular, and focus on the interface between cryptographic and human trust domains.

1 Fundamental assumptions

1.1 Overview

Blockchain technology and emergence of smart contracts made it possible to build and experiment with new forms of digital social communities and open source organisations that not only combine technical merits but can introduce economic incentives directly into the fabric of the project participants.

Since the Bitcoin experiment has started in 2008 [16], followed by the creation of Ethereum [25], enthusiasts, computer scientists and cyberpunks continouosly experiment with new autonomous systems through blockchains, decentralised finance and Decentralised Autonomous Organisations (DAOs). Decentralised Autonomous Organisation is a set of smart contracts deployed and autonomously executed on chain. This allows autonomous and verifiable computing facility that can code specific behaviour and workflows. The unifying goal of those activities is to find and design self-sovereign and self-governing systems based on cryptographic trust layer.

As of March 2022 there are over 18,000 projects that utilise the notion of self-funding, economic incentives and some form of open source software development[1]. Public blockchain ledgers offer im-

---

[1]Source: http://coinmarketcap.com

mutable verifiable registries. The diverse DAOs and decentralised applications allow contractual agreements and workflows to be coded as smart contracts[2], automated, and enforced on the protocol level.

There are multiple examples of DAO protocols that fulfil many diverse functions. For example in the decentralised finance space: stable coins and collateralised loans with MakerDAO [7], Uniswap [1] for decentralised currency exchange, or AAVE to facilitate loans on chain [12]. Despite hype and a lot of marketing jargon, the current experiments fall short in producing truly self-sovereign, decentralised and privacy-preserving systems that are capable of regulating and managing themselves in similar fashion as the contemporary centralised institutions. The systems almost always rely on centralised cloud infrastructure to provide the necessary off-chain services. The systems rely on users obtaining initial tokens or crypto funds from centralised service providers and exchanges. Majority of projects have centralised governance with a promise of "decentralisation" done later. The existing technology and experiments not only failed to create a parallel financial system that is compatible with the existing one, but they also continue to recreate a broken silo-based fragmented ecosystems of competing projects, with built-in lock-in mechanisms, and tight vertical integration that promotes winner-takes-all monopolies. This hinders innovation and creative competition.

In the following sections, we point out what the existing systems lack in order to provide governance and normative support that is capable to recreate structures and dynamics exhibited in traditional institutions. We introduce a new self-sovereign evolution of the DAO concept, called Self-sovereign Digital Institution **SDI**. SDI combines the well-formulated design elements from Open Systems Interconnection (OSI) model and the existing open protocol design from SSI standards, and bridges algorithmic trust with human-based trust in a seamless digital ecosystem that enriches the possible digital interactions between people and digitally facilitated groups. The paper concludes and argues for rebuilding the next generation of decentralised

---

[2]Without the loss of generality we treat the term `smart contract` as "code that runs autonomously in decentralised fashion on chain".

internet with the lessons learned from OSI and open interoperable protocols, combined with Web3. The enabler is decentralised identity, and we argue that $Web4 = Web3 + SSI$.

1.2 God protocols

In 1997 computer scientist Nick Szabo hypothesized a virtual computer that would provide a trusted mechanism of disintermediation between networked participants. He imagined network protocols that can be trusted by anyone, and at the same time provide unbreakable privacy and trust to all the participants. He called those cryptographic constructs "*God Protocols*" [20]. The idea was to use computer algorithms and cryptographic protocols to achieve a digital trusted third party equivalent to a trusted institution from the real physical world. The goal was to actually dis-intermediate the interaction through the use of open and friction-less cryptographic protocols. To replace third party with an algorithmic mediator, that facilitates the flow of information, transactions or value between participating individuals.

It has been envisioned, that with the use of technology, the networked participants would be able to increase the transaction throughput and eliminate frictions, lower the transaction costs and limit the potential problems with traditional, human-based intermediaries. There are multiple problems with traditional institutions and third-party human-based intermediaries, including corruption, power abuse, discrimination, and market manipulations to name a few. The most cited problems vary from various forms of inefficiencies, human errors, transaction and individual interactions costs, and physical limitations such as operating timescales (e.g. throughput). On top of that, there is also privacy concerns and centralisation issues. To address and avoid above mentioned problems, the goal is to replace human-based third parties and intermediaries with a cryptographic trust layer, that will not have the problems and limitations of traditional human-based intermediaries.

The privacy aspects of existing facilitators and intermediaries turned out to be unreliable in the digital space due to commercial self-interest and creative forms of rent-seeking behaviour (see e.g. Facebook, Google or user data resellers). Personal data tracking, profiling, and exploiting social graphs turned out to be an enormously profitable endeavour with questionable societal value. On top of that, there are many cases of abuse, data breaches, or other forms of user-data misuse. From the efficiency, security and privacy perspective, the ability to digitally dis-intermediate the interactions and to be able to replace the intermediaries with cryptographic trust layer is highly desirable. Can it be achieved today? Yes. With the innovation of cryptographic protocols, the emergence of blockchains, smart contracts and the `Web3` protocols, the original vision of automated digital *God protocols* is becoming a reality. Modern technology allows the digitisation of many aspects of our social interactions. Note, that the vision of replacing trusted intermediaries does not replace the human-trust between the participants. The actual human-trust relationships between participating stakeholders cannot, and are not intended to be entirely replaced by the cryptographic-trust. To stimulate open innovation and growth without building competing monopolies and vertical silos, and to build the ecosystem for sustainable, open innovation, the protocols must be designed with user-centric privacy in mind, interoperable, royalty free, open, fiction-less, decentralised on multiple levels of abstraction, and, provide ability to include governance framework with rules, incentives and sanctions. We will take a closer look into those criteria in 2.

1.3 Decentralised virtual computer

In 2008, an unknown group initiated a global Bitcoin [16] experiment. The idea was to use a cryptographic protocol combined with engineered consensus and incentive mechanisms to achieve global data consistency without the need for all participants to be trusted. Bitcoin is the first example of a new decentralised technology dubbed *blockchain.* By arranging the consensus mechanism as well as the economical incentives of the participants (stakeholders and users), blockchain systems can achieve global transaction data consistency and enable, in the case of systems similar to Bitcoin, a global ordered transaction ledger. This can be used as a financial instrument, a *cryptocurrency*. In subsequent years, the concept of a persistent, consistent and verifiable transaction ledger has been taken further, by the introduction of a general-purpose programming facility that mimics the envisioned Szabo's virtual computer. The first general-purpose programmable blockchain, called Ethereum, was conceived in 2013 [8], and the first release happened in July 2015. Since then, different enhancements as well as many experiments with various consensus mechanisms, governance models, as well as on-chain and off-chain services have been tried, leading to over 18,000 different publicly traded blockchain and cryptocurrency systems[3]. Besides Ethereum, multiple alternatives achieve similar properties, such for example, to name some: Cardano[3][4], Hedera [4][5], Polkadot [24][6] or Dfinity[7].

The decentralised virtual computer is like a virtual machine that is operating autonomously in decentralised fashion supported by a community of stakeholders and users. It is not directly owned or controlled by a single entity, and it is not part of any specific jurisdiction. A well-suited metaphor is to use the term *commons*. The "commons" provide computing and storage facilities that are domain, law, and jurisdiction agnostic, and provide so called *cryptographic trust layer* on top of which various projects and value added initiatives can be built. Many future stakeholders or applications can benefit from such a *common*. Good examples of such a *common* are SMTP[8] or HTTP[9] protocols. They enable rich and diverse web services that we observe today.

In order to discuss and compare various models and initiatives to build and provide such a common layer, we need a set of well defined comparison criteria, terms and concepts, thus we focus on defining the terms in the next section.

---

[3]As of March 2022, there are over 18,000 different cryptocurrencies with over 2.5 trillion USD market capitalisation

[4]`cardano.org`

[5]`hedera.com`

[6]`polkadot.network`

[7]`dfinity.org`

[8]Simple Mail Transfer Protocol, https://www.ietf.org/rfc/rfc2821.txt

[9]Hypertext Transfer Protocol, https://www.ietf.org/rfc/rfc2616.txt

2 Comparison criteria

In order to analyse and compare existing models and technologies as well as to stimulate and frame the discussion, we have to define the requirements and expectations, such that we can map existing projects and ideas along a set of well-defined dimensions. Note, that presented below metrics to be treated qualitatively, as independent dimensions. Proper formalisation and operationalisation of these metrics is beyond the scope of this concept document.

Privacy and anonymity.

User data privacy and anonymity (or pseudo-anonimity) are often reffered to as de-facto global, well-established properties of all blockchain systems. It is taken for granted that the *cryptographic trust layer* provides privacy and anonymity. This is misleading and inaccurate. Pseudo-anonymity and anonymity have varying properties and these are never permanent. The privacy and anonymity properties are not constant, but rather these are a function of time and resources necessary to deanonymise or compromise the cryptographic privacy layer. In other words, the algorithmic trust layer is always ephemeral.

The second important element is that privacy and anonymity properties are not commutative. The fact that a given hashing or encryption algorithm does not have demonstratable vulnerabilities, does not mean that an application using the algorithm is secure or privacy-preserving. When discussing or comparing systems in the context of privacy and anonymity, we need to include all abstraction layers, and include human factors, side-channel vulnerabilities, and all other factor influencing the overall system properties.

The actual value of privacy and anonymity metric for a given system is the one from the weakest link in the value chain. For example, the fact that users of Bitcoin willingly go through KYC and formal deanonymisation to obtain tokens means that many of the intended protocol protections from the lower layers of the stack (protocol level protections) are voided. KYC of some Bitcoin users weakens the privacy and anonymity of everyone using Bitcoin. We will discuss privacy and anonymity in more detail in 2.1.

Openness and interoperability.

When the entry to the system and access to its services is unrestricted and open to anyone, the system exhibits high degree of openness. If the system can be independently improved by additional layers built on top, if the system is designed without strong coupling and vertical integrations, if the system can tolerate variations in the aspects of its functioning, the system is considered interoperable. Independent implementations and value added services can co-exist and enrich the innovation and value to both, the service providers and users. Such designed systems and protocols create so called *blue ocean innovation* [11].

Both of these metrics are not easy to express formally, and, the situation is complicated by the fact that the metrics for the same system can vary on multiple levels of abstraction. For example for a given system it can be different on the service provider level vs. the users level.

IRC,[10] TCP/IP,[11] SMTP, IMAP,[12] HTTP, Tor [14], Bittorrent [26] or IPFS [5] are all protocols that are friction-less, open and interoperable on the user-level. Some of the protocols require certain degree of trust, or trust-relationships on the service level and it introduces friction and certain level of human trust requirements for core node operators. Users as well as providers can use the protocols freely, and there is no restrictions to access or flow of transactions. Protocols such as Bitcoin or Ethereum are less open and interoperable. Consider a "hard fork" of an open source project - the best of multiple forks can be combined and re-introduced into the original project or into the community, through merge requests for example. This technically is possible on the infrastructure layer for blockchain projects too, but, at the same time the economic incentives of hard forked blockchains cannot be reconciled inside the currently operating systems and hard forks. We will look closer into this topic and analyse the openness in

---

[10]Internet Relay Chat Protocol, https://www.rfc-editor.org/rfc/rfc2812.txt

[11]Transmission Control Protocol, https://www.ietf.org/rfc/rfc793.txt

[12]Internet Message Access Protocol, https://www.ietf.org/rfc/rfc2060.txt

the section below, 2.2.

Centralisation.

Centralisation and decentralisation are often misunderstood, due to the fact that there are multiple cross-cutting concerns that those properties apply to. One can consider centralisation in terms of infrastructure or in terms of governance and control. Centralisation and decentralisation can be measured on different levels of abstraction as well as on different layers of the infrastructure stack. The metric on the same system can have very different score for users, owners or programmers, infrastructure and service providers. The users and the owners can be the same or different groups, and the concepts such as: infrastructure centralisation, value centralisation, *governance* centralisation, need to be considered independently in the analysis together with other cross-cutting concerns. The statement: "the system is decentralised" is meaningless without qualification what exactly is being decentralised. Google infrastructure is highly decentralised, more than Bitcoin's, yet, Bitcoin is more decentralised in terms of governance and decision making.

The formal treatment of centralisation metric requires substantial amount of work and goes beyond the scope of this article. Here, we only highlight the main ideas qualitatively. IRC, Bittorrent, SMTP, IMAP, Tor, TCP/IP, HTTP are all examples of open, decentralised protocols for: the users, infrastructure and service providers. The governance and ownership however are already a more complicated matter, even for those relatively simple and well defined protocols. SSL and HTTPS, rely on Certification Authorities (CAs), that are centrally managed, but implemented in a pluralistic and decentralised fashion. SSL and HTTPS are open standards and royalty free protocols, and this allows competition and emergence of free services providers, e.g. *Let's Encrypt*[13], that offer privacy protection and the services for free to the end-users. This type of pluralistic offerings is not possible for example in existing blockchain systems that have payements and fees introduced on the infrastructure level. This is equivalent to HTTPS charging fees on the protocol level and giving

---

[13]Let's Encrypt, https://letsencrypt.org/

the fees to CAs for the service they offer.

Even though the above protocols are designed well, stimulate open innovation and ability to build value added services on top in a frictionless fashion, all of them were designed in a somewhat centrally controlled process. In some cases the ownership is centralised through consortia or expert working groups (some having more open and friction-less access policies).

When discussing centralisation and decentralisation, one needs to qualify exactly what level a given measure is applied to, and what is being covered. For example, Bitcoin is quite well decentralised in terms of ownership, governance, or infrastructure, however, the value distribution in the network is highly centralised [19]. Because of the inherent complexity of the issue related to decentralision, we argue that specific cross-cutting concern must be used as a context. Later in the article we discuss decentralisation in the context of infrastructure (3.1), in the context of governance and decision making (2.3), and in the context of normative behaviour and sanctions (2.4).

Friction.

Any built-in mechanism that slows down or in any way inhibits the interactions can be considered as introducing friction. In many situations friction is introduced by design, e.g. speed limits on highways reduce severity of accidents and make the system overall more reliable and efficient. HashCash [2] introduces friction to SMTP to prevent abuse and spam. And Proof-of-Work introduces friction in order to secure the ledger and achieve the consensus. Friction can and often is used as an abuse prevention mechanism. Well designed, desirable friction should affect abuse cases, but not normal non-abuse system usage. In some situations the friction is an undesirable side-effect of design decisions. When friction affects normal use cases (not abuse use cases) it can be considered undesirable or even harmful. Mechanisms that limit system throughput, mechanisms that introduce fees, and mechanisms that limit participation are all introducing friction (some desirable or some undesirable). IRC, TCP/IP, SMTP, IMAP, HTTP, Tor, Bittorrent or IPFS in normal use cases do not introduce any friction. Bitcoin introduces friction through the proof-of-work and by the internal clock of the blocks, which is by design limited to tick at approximately 10min interval. Transactions have fees, which is also an element of friction. Ethereum introduces similar frictions to prevent abuse on the Ethereum Virtual Machine (EVM) level. Due to the nature of smart contracts, there are additional restrictions on the complexity of the smart contracts and the amount of data stored on-chain, which can also be considered built-in frictions on the protocol level. Protocols such as Binance Chain [6] reduce some of the frictions by introducing centralisation. It is important to note that centralisation often counteracts necessary frictions, and, some of the decentralisation solutions and decentralised coordination requirements inherently need to introduce friction.

Transparency and the ability to self-correct.

For autonomous systems to be able to self-correct, adjust and prevent abuse, the system needs to exhibit self-reflection and the ability to reify its own operation and design parameters. For this to be possible, the system must expose essential metrics and properties in a transparent and verifiable fashion. The degree to which the system is transparent is a proxy of how much self-correcting reification the system can exercise.

RAI vs RAW. Expression and enforcement of the internal rules and laws.

This point intersects the algorithmic-trust and human-trust layers and it is about the interface between both of those worlds. The internal rules and laws always have two aspects: rules as written (RAW), and rules as intended (RAI). On the algorithmic trust layer it is possible to achieve a good mapping between RAW and enforcement done on the protocol level itself.

The DAO experiment [15], as well as many problems within the contemporary Decentralised Finance, demonstrate that it is not possible to achieve a perfect mapping between RAI and RAW. Systems that provide resolution mechanisms and systems that focus on RAI and improvements of RAW to fit better and better of the RAI, are better, more adaptable, agile, and future-proofed compared to systems that lack such mechanisms of mapping RAI into RAW. Most noticable example of RAW-based system where RAI is almost never really consider is Ethereum, in which smart contracts are immutable and any bugs or problems with coding of the RAI into RAW (smart contracts) result in systemic failure, funds loss, or other forms of misalignment. The next generation protocols and infrastructures to be used for decentralised finance must take into account the RAI vs RAW dillema.

2.1 Privacy

When discussing privacy and anonymity it is hard to escape ideological arguments. Some use notions that allude to anonymity and privacy in idealised, absolute terms, e.g.: "Bitcoin is anonymous money."[14] There are two important points to cover. One, that privacy and anonymity are always transient, ephemeral. One can be anonymous or retain data privacy on the technology layer only for a certain, finite duration of time. Privacy and anonymity are not features that can be maintained forever. This is summarised well in the "Attacker advantage"[10] as well the "Rubber-hose cryptoanalysis"[15] metaphor. Note, we use the rubber-hose argument broadly, not limited to physical coercion - it can include any form of coercion, social engineering or deception techniques.

To maintain anonymity (or privacy), one needs to use perfect, vulnerability-free software, all the time, without making a single mistake, or side-channel leak of information, ever. The attacker on the other hand, given sufficient amount of resources, can wait for a vulnerability, for a mistake or side-channel leak. The attacker only needs one mistake to deanonymise or to obtain private information. Therefore, privacy and anonymity are always transient, and should be treated as temporary shelters that can protect user data or privacy, but only for a limited duration of time. Permanent anonymity or permanent privacy is an illusion. Systems that do not take the temporal nature of privacy and anonymity into consideration are flawed.

---

[14]Robinson, T. "Bitcoin is not anonymous". https://www.elliptic.co/blog/bitcoin-transactions-money-laundering

[15]https://xkcd.com/538/

The second important point is the paradox of crypto currencies, that on the infrastructure level, on the level of the P2P network and nodes, assumes anonymity, but for the service layer, the actual payment, and exchange of value, lift this requirement and have no protections against centralised service providers, deanonymisation services and blockchain crawlers, managed wallets and exchanges that require all users to voluntarily deanonymise themselves through centralised KYC procedures. This makes no sense and truly is a puzzling paradox between the marketing and the perceived anonymity and trust offered by the algorithmic trust layer and the reality.

It is exactly on the application and service level where user-centric privacy is highly desirable, whereas on the infrastructure level privacy is not required and often undesirable (to prevent various forms of Sybil attacks [23, 9]). Users should be in control of their own identity and how it is being used [22], whereas infrastructure providers and services providers do not need to be anonymous. We argue, that it would be much more desirable to reverse the model: on the infrastructure level the requirement of anonymity is not desirable, because robust abuse prevention mechanisms, inability to track and penetrate the network, are hard or impossible to achieve in efficient and friction-less protocols. Infrastructure provider anonymity introduces really hard technical challenges and problems that are currently solved by either energy or capital inefficient processes (proof of work), or through capital lock-in mechanisms that promote rent-seeking behaviours.

Lifting the anonymity on the infrastructure level would allow efficient and robust protocol designs without artificial added frictions. Paradoxically, lifting anonymity requirements on the infrastructure level would make the systems more secure and anonymous on the application and service layers. Making the service providers accountable and verifiable would prevent many abuse scenarios happening today in blockchain systems (for example, to name some: harvesting and crawling user data, de-anonimisation, price manipulation, double-spend attacks, infrastructure attacks, spam, and so on).

2.2 Open and Interoperable

Since the inception of public permissionless blockchain systems, all contemporary cryptocurrency systems suffer from various forms of centralisation and friction. Some centralisation mechanisms are enforced by the financial and governmental compliance and regulatory frameworks, which are based on centralised custodian models of financial services. The law does not consider an ability for a financial services being provided by non-custodian models. As such, those regulatory mechanism are designed to regulate central hubs such as exchanges or wallet providers. On the other hand, the currencies themselves make no distinction between an individual user account and aggregated exchange or pooled account for example. This provides an opportunity for centralisation dynamics to occur, such as mining power centralisation, proof-of-stake centralisation, or liquidity centralisation with the "winner takes all" monopoly-based markets. Bitcoin wealth distribution, measured by Gini coefficient, is one of the worst compared to GDPs of normal countries or known social systems [17].

Transaction fees and proof-of-work represent built-in frictions introduced in order to safeguard the system against abuse and sybil attacks under the assumption of participants anonymity. The problem lies in the fact that the mechanisms are built-in into the protocol itself spanning many vertical layers, and the design is a single monolith that is difficult to innovate or improve over time. Consider email protocol such as SMTP to have fees associated with it on the protocol level, and built-in mechanism for delivering emails to always have 10min delay. Service level innovation such as fee-less offering by some mail providers which monetize on user's data and targeted advertising, innovation done by Protonmail that charges users fees for value added, in their case, privacy, and free mail services for university students and employees (no fees) would not be possible to happen. Tight coupling and monolithic, vertical integration is an undersirable property of communication protocols, and does not play well with a notion of open, extensible, innovative and agile systems.

The contemporary blockchain systems are monolithic, and updates or improvements must be done in non-interoperable fashion due to the lack of proper layering and partitioning of the core protocol design. Consider Binannce chain and Ethereum Foundation chain which are not interoperable with one another, even though they are fundamentally the same (Binance chain is a fork of Ethereum). This is analogous to Google Gmail, Yahoo mail and Microsoft Outlook mail all using the same implementation of SMTP at their core, but all making sure that each is a silo with user lock-in built-in, such that sending emails from Gmail to Outlook requires "wrapping" and "cross-chain/cross-protocol gateways". For a business person this types of lock-in and integration might make sense, especially for rent-seeking. However, for users, for engineers and innovators, this is highly undersirable.

The vertical silos make sense from the silo owner's perspective as the silo provides monopoly, control, makes it harder or costly for user to migrate to a competitor, and it allows certain rent-seeking practices. However it makes little sense from the service providers' or users' perspectives. The open SMTP protocol prevents direct lock-in or rent-seeking, and makes an open levelled playing field for all possible email providers to compete and offer value added to users and service providers, without locking users into monopolies. Next generation blockchain systems have to look back to the roots of Internet and Internet protocols, take inspiration, and be re-designed without the built-in lock-in mechanisms or any form of vertical integration.

2.3 Governance

Decentralisation: Failure and abuse tolerance represents a measure of robustness to failure. Typically, it is quantified as a percentage of participants that need to be malicious, corrupt, attacked or taken down for the system to cease functioning. The perfect score is 100%. Bitcoin protocol functions up to 51% (hash power, not nodes), and some Byzantine fault tolerance (BFT) protocols can cope with up to about 70% nodes [18]. The metric is complicated due to the different levels at which it can be applied to a system, which we see in the above example because in the context of Bitcoin we have to account for Proof of Work (hash power), not only to the node count, like we did in the case of BFT protocols. Binance chain or Polygon have a small number of highly trusted and robust nodes, whereas Dash has 1000 trusted nodes. Bitcoin or Bittorent operate in large volume but do not have the concept of "trusted nodes" built-in. Correct comparisons between various design can only be qualitative and also limited. Many P2P protocols are designed to achieve a perfect score. BitTorrent and some of the Distributed Hash Tables (DHTs)[16] provide examples of protocols with a perfect score of 100%. As long as at least 1 node remains operational in the network, the network will continue to function properly. Note, that failure and abuse tolerance is different to sanctions and ability to exercise internal rule compliance, which we discuss next.

Decentralisation: Decision making.

Ability to make and execute decisions in a decentralised fashion is a complex, diverse and difficult topic. Most contemporary systems achieve it by de facto combining technological solutions (enforcements on the protocol level) together with social agreements and negotiations with the market and core stakeholders, like for example in the case of Bitcoin hard and soft-forks.

Decentralisation: Law creation and Law enforcement.

The ability to create rules and norms, and the ability to punish or exercise sanctions for not following the rules or norms, is fundamental for creation and maintenance of Self-sovereign Digital Institutions. On the protocol level, it is possible to enforce some behaviours or automate certain rules however, any emergent behaviours cannot be predicted up-front, and, the system must be robust and agile enough to adjust and deal with unforeseen cases. We have highlighted earlier the difference between Rules as Written (RAW) versus Rules as Intended (RAI). The intention of the rules, the meta level and constitutional objectives should be always attached to the RAWs, such that the interpretation and abuse can be detected and corrected. "Code as Law" is a great metaphor, as long as the Code reflects well RAI. However, perfect alignment between RAW (Code) and RAI is hard. It requires

---

[16]Distributed Hash Tables,
https://www.ietf.org/proceedings/65/slides/plenaryt-2.pdf

iterative process, adjustments, and continous feedback. Therefore, algorithmically expressing RAW is not enough. RAI needs to be taken into account too, and that needs to be formally expressed and attached to the Code (RAW). The community decisions in relation to norms and sanctions must go beyond the protocol level designs. This has been one of the early lessons learned due to The DAO fiasco [15].

2.4 Code, Markets and Norms

Szabo's work has been extended by Lessig's four pillar model [13]: Code, Law, Norms, and Market. The code represents the physical as well as protocol level constraints. *Law* are all the norms that are enforced by the government, whereas *Norms* represent norms enforced by the institution or social contracts. Finally, the *Market* represents economic incentives and drivers.

There are many parallels between digital and physical institutions, however, there are also many differences. Digital institutions and communities can span multiple physical jurisdictions, they can maintain a high level of privacy as well as security and they can facilitate improved transaction and interaction throughput, and scales, unlimited by constrained of the physical institutions.

3 Code

3.1 Decentralised architectures

The core fundamental element needed for peer-to-peer system to operate is the so called *gossip protocol*. Without the loss of generality, let us assume it is a mechanism for peer (nodes) to be able to broadcast messages and to talk to one another. The contemporary blockchains and decentralised finance systems are using single, fixed and global gossip protocol, often with a single fixed rendezvous point, see Figure 1. This can be contrasted with a better, more robust protocol that allows various transports and proper decentralisation of communication, see Figure 2.

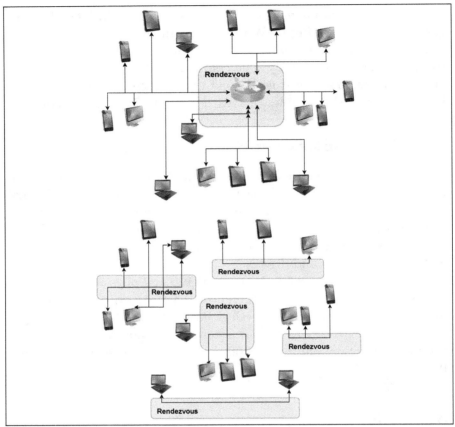

Figure 2: Decentralised communication, e.g. DIDcomm.

The fundamental infrastructure layers should use the principles and experiences gained from Open Systems Interconnect (OSI) work and aim at separating the various communication layers into a stack, such as to aid interoperability. See 3.2 for details. Instead of building vertical siloed designs that promote walled gardens and are extremely hard to provide interoperability, contemporary DeFi systems should take inspiration from OSI and design the systems with interoperability as a main goal. This is what has worked well for SSI community in order to break down the siloes created for managing and exchanging digital identity data.

Contemporary defi systems almost always use `libp2p`, which offers great foundation for interoperable sevices, but, they do that in such a way as to inhibit, or completely disallow interoperability. See **??** for details.

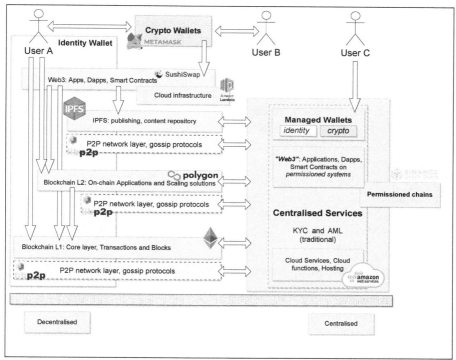

Figure 3: User interaction with decentralised applications.

What it means is, that node software for a particular application cannot re-use existing node code from another application, but, needs to bake everything afresh. It is similar to statically linked libraries versus dynamic libraries. Ideally, it should be possible for the same node executable, to be dynamically re-configured to support various protocols and applications, given, that almost all of them fundamentally are based on libp2p anyway. There is no technological reason to create those walled gardens.

The situation is even worse, given for example Lightning Network, which is an application layer on top of bitcoin, redoing new gossip protocol and new completely independent network layer with nodes, even if, each Lightning Network node must be by design also a Bitcoin node proper. A simple ability to extend a bitcoin node to also support gossip on the application layer for Lightning Network would be a reasonable technical approach, and users would instruct they node clients to either enable or not, specific applications build on top of Bitcoin. However, for business and ownership reasons, the two, tightly interlinked projects maintain independence and create their own walled gardens. This, for a technologist, is similar, that everything needs to be based on forked code bases, and always statically linked into the final executable - that makes it hard to inter-operate, collaborate and compete on improvements.

A better approach would be to define open, public protocols (interfaces) and data schemas, and allow people to build business logic and application on top of those public protocols/interfaces. Similar discussions occured in the early stages of the Internet, where OSI, IETF and W3C helped in making the Internet what it is today.

3.2 Open Systems Interconnection

The Open Systems Interconnect has designed a model for breaking down communication into layers such that value-added services can be added within and on top of various layers. The generic scheme is provided in Figure 4. Note, in networks a design of specific inter-layer protocols has been identified, and some of the internet protocols do not sit exactly at the OSI layers. The most widely used examples are MPLS (Multi-protocol Label Switching) and PPPoE (Point-to-Point Protocol over Ethernet) both of which are examples of a so called Layer 2.5 (a layer in between Layer 2 and Layer 3). This is understandable and often desirable, to extend the OSI model into more fine grained layers to enable robust applications and patch existing protocol stack to avoid centralisation and vertical lock in. Note, it is better to use re-usable MPLS across a broad range of protocols rather than build routing solutions that are baked in on Layer 2 and Layer

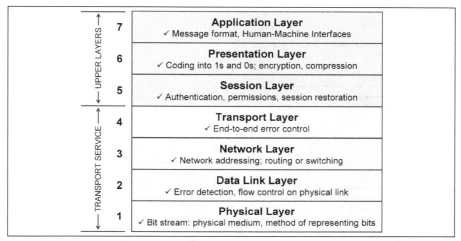

Figure 4: OSI layers (credit CISCO training materials).

3, which would hinder extensibility and ability to build value added on top of Layer 2.5.

In Figure 4 the original OSI model is depicted. In Figure 5 a new proposed protocol stack that bridges various communication layers, as well as the cryptographic/algorithmic trust and human-based trust are bridged. In the figure, the candidate for the networking layer is depicted as DIDcomm and DID infrastructure. This is because at the time of writing, DIDcomm provides the most robust and transport independent methods of peers to communicate with one another, and there are ongoing attempts to bridge `libp2p` and DIDcomm [21].

3.3 Self-sovereign identity

The Internet protocols, similarly to cryptocurrency protocols, do not include the concept of identity or verifiable credentials by design. Those aspects were left out of the original specifications. Those aspects are something that needed to be added to enable fainance, commerce and other broad range of services in the digital world that rely on some form of identification, or credential verification. Let us consider the most fundamental one, based on a key of username

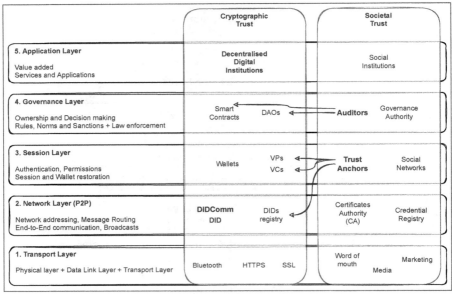

Figure 5: Layered SDI architecture.

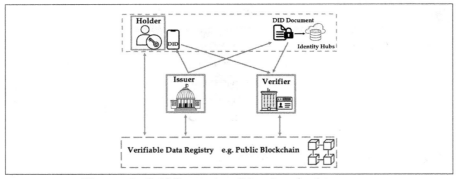

Figure 6: Self-Sovereign Identity Architecture.

and password (or email address and a password). In that model, the particular system controls the "identity" by storing the identifier as well as means of verifying the credential. Note, that those various solutions continue to be based on aggregation within walled gardens, as one needs a different username-password pair for different Internet service. To enable sharing the same username-password pair for various services, ie. to decouple the identification and credential verification with a form of delegation, one could extend the basic model to OAuth or OpenID. That however continues to rely on a centralised controller that controls and stores the pairs of users in a central place. If that provider locks the user out, the system stops working for that user (if a user of Facebook uses Facebook login to access third party service and Facebook is not online or deletes the user, the user loses access to all services for which Facebook-controlled ID was used).

Those electronic systems mimic the physical methods of identification and verification of credentials, with the issuers controlling the identifiers, credentials and the actual protocols of how the identifiers and credentials are being used. In the above example, users using Facebook login need Facebook to authorise the credentials every time a 3rd party requires that.

The users' experience, privacy, and security suffer in the above models. The existing identity systems allow personal tracking, misuse of personal privacy and other forms of digital identity abuse. The traditional solutions, often based on a siloed, walled garden approach, result in privacy, security and interoperability issues. One obvious problem is the so-called call home problem, which is the inability to efficiently allow offline verification of credentials (and identity) in cases where the verifier is not available online (e.g. due to outage or denial of service attack). A service relying on an aggregator is not able to verify the truth of data unless they connect home to the aggregator and get a successful reply. Such an aggregator knows where this data was used, who used this data and when. That is a powerful profiling mechanism. The aggregator model has a big problem with privacy, lock-in effects and monopolistic ways to handle data.

The concept of Self-Sovereign Identity (SSI) is that each user fully controls their own information. Users can add, remove and share attributes at their own discretion. They can share their email to a service provider and then subsequently revoke the rights to use this email. There are even good privacy solutions that make the use of selective disclosure, so that you are only sharing the data attributes needed for that particular exchange. Moreover, users can have one or more identifiers (something that enables a subject to be discovered and identified) and can present claims relating to those identifiers without having to go through an intermediary. Claims made about a user can be self-asserted or asserted by a third party whose authenticity can be independently verified by a relying party. Under self-sovereign identity model, identities must not be held by a single third-party entity.

SSI solutions are based on open standards and protocols like verifiable credentials (VCs)[17] and decentralised identifiers (DIDs)[18] can enable bridging contexts. VCs are containers capable of carrying any sort of verifiable data payload. The payload of a VC can be any kind of data and is not limited to a credential. DIDs are a new type of identifier that enables the controller of a DID to prove control over it without requiring permission from any other party. VCs with DIDs can bring transitive trust with rapid verifiability.

Users are in control of their data and can offer-accept connections to-from other people, organizations, or things and establish peer-to-peer relationships. They can store offered VCs in their digital SSI wallet and share them with verifiers as shown in Figure 1. Verifiers can confirm cryptographically four things (who issued it; it was issued to the user and not someone else; it has not been tampered with; it has not been revoked by the issuer) about what the user has shared, without having to contact the original issuer. In such an ecosystem, users act as couriers of verifiable data between organisations they deal with and unrelated trust domains are bridged without requiring direct connections to one another. The blockchain could act as a *verifiable data registry* and replace the centralised registration authority in traditional identity management systems.

---

[17]W3C Verifiable Credentials Data Model `https://www.w3.org/TR/vc-data-model/`

[18]W3C Decentralized Identifiers v1.0 `https://www.w3.org/TR/did-core/`

3.4 Code as Norms

Arguably by design, contemporary cryptocurrencies do not have built-in mechanisms to differentiate users from different jurisdictions, individual users from commercial stakeholders, or an ability for a user to provide proof of their own identity. Even though those properties are necessary for working financial systems, those are subsequently enforced on the application layer through Know-Your-Customers (KYC) and Anti-Money-Laundering (AML) regulations exercised by the centralised custodian service providers. Note, that KYC and AML are de-facto not designed for custodian services only, but actually operate in peer-to-peer fashion through cash-like systems, banks, and some high-value shops (e.g. purchase of a vehicle or house in many jurisdictions requires payment conducted via bank transfer, and cannot be done by cash payment).

Some of the problems in the current DeFi services are rooted in the inability to align RAWs to be representative of RAIs. This can happen due to unintentional bugs, inability to formally express all the rules, or through unintended consequences of RAWs. There are differences between internal norms and laws. By laws, we refer to RAIs that have been encoded as RAWs. By norms, we refer to RAIs that have not been codified. Coding all RAIs strictly as RAWs is not possible, and not done in normal physical institutions neither. In contemporary systems, most of the RAIs are written in a form of white papers or technical documents, and they only work as guidance. No enforcement of those RAIs can be conducted, and in fact, there is a clear disregard towards RAIs in favour of bugs and errors in RAWs. "Code as Law" is a common phrase among contemporary DeFi enthusiasts. The disregard for RAIs is puzzling when observed from outside of the communities, but has been broadly accepted as "normal" in the current DeFi communities.

The ability to encode RAIs in formal documents and the ability to to resolve conflicts in favour of RAI instead of RAW is one of the main concepts behind SDI. This is inspired by some of the countries, e.g. Norway in which legal system stresses the need and the precedence of RAI, not RAW. If there is a mismatch between RAI and RAW, RAI always trumps RAW and RAW is updated to better reflect the RAI.

Property	A	B	DAO	SDI
Privacy	not supported			application layer
Transparency	partially supported			supported
Anonymity	pseudonyms			application layer
Identity	not supported on the protocol, but supported on the application level and via KYC on centralised services			supported on all layers
Openness	low	low	medium	high
Interoperability	very low	low	medium	high
Friction	high	high	high	low on protocol level
RAW	very limited	medium	high	high
RAI	not supported/only informal			formal
Governance	low	low	medium (only RAW)	high (RAI & RAW)
Self-sovereign	low	low	medium (lack of law enforcement)	high (law-enforcement built-in)

Table 1: Feature comparison between A (generation 1 blockchains, e.g. Bitcoin), B (generation 2 networks e.g. Ethereum, Polkadot, Cardano and scaling solutions), DAO (smart-contract-based DAOs and DeFi protocols, e.g. Uniswap), and **SDI**.

4 Discussion and Future work

4.1 Summary

In order to provide digital institutions that can compete and supplement those evolved in physical world the designs of new digital decentralised systems need to (re-)use open, interoperable standards, which are designed and inspired by the OSI model, incorporate consensus and value distribution mechanisms that promote decentralisation and information symmetry, and avoid mechanisms that enable single-player advantage or ability for group creation and centralisation. The new designs must implement governance, both in terms of rules and sanctions, such that the system can adjust and evolve to avoid new, unforeseen initially threats and threat actors, and to be able to maintain robustness and scalability via well-established mechanisms for digital systems evolution. Protocol improvements should be facilitated by layered and decoupled architectures, that promote open innovation and gradual improvements. Value added services and value creation should be restricted to application and service layers, making all the underlying stack into an open platform (a common), that can be freely shared and cared for, and contributed to, diverse stakeholders.

4.2 Future work

The conceptual model presented in this article provides the initial framework for discussing and building interoperable open protocols that can enable the envisioned decentralised Self-sovereign Digital Institutions. The actual architecture and the layered model is incomplete. The bootstraping of trust is not solved and intermediate solutions based on existing trust anchors might need to be employed. The governance frameworks even on the SSI level are also work in progress. There is no clear unified solution for decentralised credentials, or robust trust anchors that could work in a fully decentralised fashion at the moment. All of which are active area of research and development, involving many startups, standardisation efforts, eg. W3C or

Decentralised Identity Foundations. DIDcomm offers robust Layer 2 in our SDI layered model, but working out the actual details of the needed DID methods is left as future work.

References

[1] H. Adams, N. Zinsmeister, M. Salem, R. Keefer, and D. Robinson. Uniswap v3 core. Technical report, Technical report, 2021.

[2] A. Back et al. Hashcash-a denial of service counter-measure. 2002.

[3] C. Badertscher, P. Gaži, A. Kiayias, A. Russell, and V. Zikas. Ouroboros genesis: Composable proof-of-stake blockchains with dynamic availability. In *Proceedings of the 2018 ACM SIGSAC Conference on Computer and Communications Security*, pages 913–930, 2018.

[4] L. Baird, M. Harmon, and P. Madsen. Hedera: A public hashgraph network & governing council. *White Paper*, 1, 2019.

[5] J. Benet. Ipfs-content addressed, versioned, p2p file system. *arXiv preprint arXiv:1407.3561*, 2014.

[6] Binance. Binance chain documentation. Technical report, Technical report, 2022. Available at `https://docs.binance.org/`.

[7] M. Brennecke, T. Guggenberger, B. Schellinger, and N. Urbach. The de-central bank in decentralized finance: A case study of makerdao. In *Proceedings of the 55th Annual Hawaii International Conference on System Sciences*, 2022.

[8] V. Buterin, 2013. Available at `https://ethereum.org/en/whitepaper/`, Accessed 19 April 2022.

[9] W. Casey, A. Kellner, P. Memarmoshrefi, J. A. Morales, and B. Mishra. Deception, identity, and security: the game theory of sybil attacks. *Communications of the ACM*, 62(1):85–93, 2018.

[10] C. M. Christensen and R. S. Rosenbloom. Explaining the attacker's advantage: Technological paradigms, organizational dynamics, and the value network. *Research policy*, 24(2):233–257, 1995.

[11] W. C. Kim and R. Mauborgne. Value innovation: a leap into the blue ocean. *Journal of business strategy*, 2005.

[12] S. Kulechov. AAVE Protocol. Technical report, Protocol Whitepaper, 2020. Available at `https://github.com/aave/aave-protocol/blob/master/docs/Aave_Protocol_Whitepaper_v1_0.pdf`.

[13] L. Lessig. *Code: And other laws of cyberspace*. Basic Books, December 2006. 2nd Revised Edition.

[14] D. McCoy, K. Bauer, D. Grunwald, T. Kohno, and D. Sicker. Shining light in dark places: Understanding the tor network. In *International symposium on privacy enhancing technologies symposium*, pages 63–76. Springer, 2008.

[15] M. I. Mehar, C. L. Shier, A. Giambattista, E. Gong, G. Fletcher, R. Sanayhie, H. M. Kim, and M. Laskowski. Understanding a revolutionary and flawed grand experiment in blockchain: the dao attack. *Journal of Cases on Information Technology (JCIT)*, 21(1):19–32, 2019.

[16] S. Nakamoto. *Bitcoin: A Peer-to-Peer Electronic Cash System*. Bitcoin.org, 2017. Available at `https://bitcoin.org/bitcoin.pdf`, Accessed 19 April 2022.

[17] A. R. Sai, J. Buckley, and A. Le Gear. Characterizing wealth inequality in cryptocurrencies. 2021.

[18] G. Shapiro, C. Natoli, and V. Gramoli. The performance of byzantine fault tolerant blockchains. In *2020 IEEE 19th International Symposium on Network Computing and Applications (NCA)*, pages 1–8. IEEE, 2020.

[19] B. Suisse, 2019. Available at `https://www.bitcoinsuisse.com/research/decrypt/distribution-of-stores-of-value`, Accessed 3 May 2022.

[20] N. Szabo. The god protocols. *https://nakamotoinstitute.org/the-god-protocols*, 1997.

[21] O. Terbu and A. Horvat. Didcomm messaging through libp2p. `https://medium.com/uport/didcomm-messaging-through-libp2p-cffe0f06a062`, Nov. 29 2021.

[22] A. Tobin and D. Reed. The inevitable rise of self-sovereign identity. *The Sovrin Foundation*, 29(2016), 2016.

[23] A. Vasudeva and M. Sood. Survey on sybil attack defense mechanisms in wireless ad hoc networks. *Journal of Network and Computer Applications*, 120:78–118, 2018.

[24] G. Wood. Polkadot: Vision for a heterogeneous multi-chain framework. *White Paper*, 21, 2016.

[25] G. Wood et al. Ethereum: A secure decentralised generalised transaction ledger. *Ethereum project yellow paper*, 151(2014):1–32, 2014.

[26] C. Zhang, P. Dhungel, D. Wu, and K. W. Ross. Unraveling the bit-

torrent ecosystem. *IEEE Transactions on Parallel and Distributed Systems*, 22(7):1164–1177, 2010.

The Cost Function of a Two-Level Inventory System with Identical Retailers Benefiting from Information Sharing

Amir Hosein Afshar Sedigh
Behpardaz Hamrah Samaneh Aval, Tehran, Iran

Rasoul Haji
University of Science and Culture, Tehran, Iran

Seyed Mehdi Sajadifar
University of Science and Culture, Tehran, Iran

Abstract

This paper investigates a two-echelon inventory system with a central warehouse and $N(N > 2)$ retailers managed by centralised information sharing mechanisms. In particular, the paper mathematically models an easy to implement inventory control system that facilitates making use of information. Some assumptions of the paper include: a) constant delivery time for retailers and the central warehouse and b) Poisson demand with identical rates for retailers. The inventory policy comprises of continuous review (R, Q) policy on part of retailers and triggering the system with m batches (of a given size Q) at the central warehouse. Besides, the central warehouse monitors retailers' inventory and may order batches sooner than retailers' reorder point, when their inventory position reaches $R + s$. An earlier study proposed the policy and its cost function approximation. This paper derives an exact mathematical model of the cost function for the aforementioned problem.

Keywords: Two-echelon inventory; Information sharing; Poisson demand; Continuous review.

1 Introduction

Improving flexibility and approachability is essential in uncertain and competitive markets. As such, supply chains (SCs) have put more concentration on information technology to distribute benefits among the members [40]. No information sharing incentivizes individual members to maximize their own benefit. Therefore, customer demand variability is strengthened in upstream levels (e.g. manufacturers and vendors). This leads to a phenomenon called the bullwhip effect, originally reported by Forrester [14] as "demand amplification". The bullwhip effect was studied by many researchers, such as Simchi-Levi et al. [34]) and Li & Simchi-Levi [24]. Not only information sharing in a centralized decision-making system weakens demand variability in upstream levels, but it also has benefits such as a fair distribution of profits among all members [25].

Several researchers experimentally and theoretically studied benefits of information sharing in SCs. For instance, Cho & Lee [11] studied a seasonal SC including one supplier and one retailer. The study indicated that information sharing is beneficial if the lead time is shorter than the seasonal period. Liu et al. [26] examined the impact of information sharing and process coordination on logistic outsourcing in China. This study showed that these integration mechanisms, especially information sharing, are beneficial. Chen et al. [10] found full information sharing scenario as the best scenario by investigating usage of eight scenarios regarding sharing different types of information. Janaki et al, [21] indicated the value of information sharing for improving SC performance under uncertainty. However, Ojha et al [30] indicated the importance of careful choice of type of shared information, when some types of information are not shared for managerial concerns. Liu et al. [27] indicated that sharing basic inventory information, disregarding production capacity and resource constraints, has the highest impact on the coordination in a decentralized supply chain.

Li & Wang [25] investigated such problems under three policies, namely installation stock, echelon stock, and information sharing.

The study indicated that installation stock policies are easy to operate but neglect performance optimization for not utilizing information about the customers – see Axsäter [2, 3, 4], Forsberg [15, 16], Simchi-Levi & Zhao [35, 36] for various approaches of using installation stock policies. In echelon stock policies, information on the cumulative inventory positions of all downstream installation is available. Chen & Zheng [8, 9] evaluated (R, nQ) echelon stock policies in serial inventory systems and its extension for a multistage inventory system with compound Poisson demand. For the latter they evaluated optimum boundaries and proposed an algorithm for near optimal solution that can also be used for exact solution. It is worth noting that Axsäter & Rosling [6] indicated that echelon stock policy may outperform the installation stock policy.

In information sharing policy, all or part of information about the inventory position, bill of materials (BOM), etc. is shared among members. This can enhance SC performance due to the reduction of the forecast errors. The information sharing not only helps supply managers to secure transparency and accessibility of information, but it also facilitates decision-making by providing more accurate information about the chain [34].

In the light of these studies, some studies proposed policies that benefited from shared information. Here we state policies for two-level SC with stochastic demands. For instance, Moinzadeh's [28] policy is an easy to implement information sharing-based approach for a two-echelon SC with a product, a central warehouse, and some identical retailers under a stationary random demand. He derived an approximate cost function and proposed a heuristic optimization procedure to obtain a near-optimal policy. However, the optimization procedure could not identify the optimal boundaries. Sajadifar & Haji [31] derived an exact cost function for Moinzadeh's [28] policy, considering one retailer.

They used Axsäter's [2] installation stock model for one-for-one policy, to evaluate the cost function. Later, Haji & Sajadifar [20] derived the optimal boundaries for their proposed model in order to have less computational efforts in obtaining the optimal solution. A further development is due to Axsäter & Marklund [5] who proposed a dynamic information sharing policy for non-identical retailers, that is hard to implement because of its non-static nature. More recently, Afshar Sedigh et al. [1] derived the exact cost function and optimal boundaries of Moinzadeh's [28] policy, for two retailers.

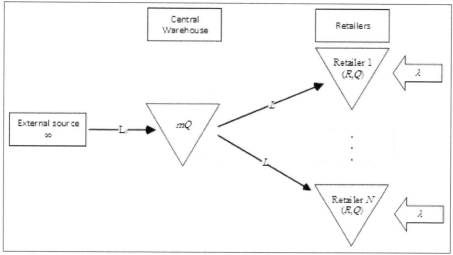

Figure 1: System Schema

This paper re-investigates Moinzadeh's [28] policy by obtaining the exact cost function and optimal boundaries for more than two retailers. This study similar to Moinzadeh [28] considers some retailers and a central warehouse who have constant delivery times. Retailers face independent Poisson demand with identical rates. The continuous review (R, Q)-policy is applied by the retailers. The system triggers with m batches of a given size Q at the central warehouse. When a retailer's inventory position reaches $R + S$, the central warehouse places an order to an outside supplier. This paper derives the exact cost function for different retailers, based on which the optimal policy can be determine. The system is split into some simpler states by conditioning. Then, the exact cost function is obtained for each state. A brief schema of the presented problem is illustrated in Figure 1, where the notations immediately follow in Section 2.

2 Problem definition and notations

This section presents employed notations and assumptions. Note that we utilize similar notation to earlier studies for ease of researchers familiar with the field.

2.1 Notations

The notations used in this paper are as follows:

L: Delivery time from the central warehouse to each retailer

L_0: Delivery time from the external supplier to the central warehouse

λ: Demand intensity at each retailer

h: Holding cost per unit per unit time for each retailer

h_0: Holding cost per unit per unit time at the central warehouse

β: Shortage cost per unit per unit time for each retailer

Q: Given system's order quantity

R: Reorder point for each retailer

m: Number of initial batches allocated to the central warehouse

$A_0(t)$: Set of retailers whose inventory positions are less than or equal to $R + s$ at t

$A_1(t)$: Set of retailers whose inventory positions are more than $R + s$ at t

Moreover, similar to Axsäter [2], the following notations are used:

$\gamma(S_0)$: Average holding cost in the central warehouse per unit when the inventory position at warehouse is S_0

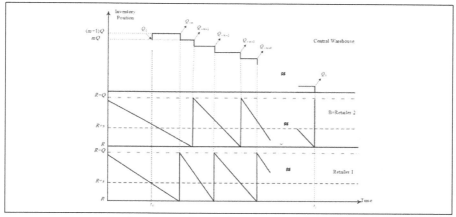

Figure 2: An example of the system for two retailers

$\pi^S(S_0)$: Average holding and shortage costs in a retailer when inventory positions of the central warehouse and the retailers are S_0 and S, respectively.

Figure 2 represents the notations, note that for simplicity in our figures. We use a continuous representation of demands to the retailers. Note that the $[m + 1]^{th}$ batch is Q_0; therefore, we use negative numbers for the m batches received before this time (i.e. $-m$ to -1). The rest of the notations are as follows:

Q_0:   An arbitrary batch of size Q, chosen for investigation

t_0:   Ordering time of Q_0 by the central warehouse

t_1:   Ordering time of Q_0 by one of the retailers

B: The retailer who orders Q_0

t^+:   A moment just before t

t^-:   A moment just after t

k: Customer demand between t_0 and t_1^+

177

j: An arbitrary unit of Q_0

$TC(m, R, s)$: Average system cost per unit time

In what follows, we state the assumptions and inventory policy of this paper.

2.2 Assumptions

Similar to Moinzadeh [28] and Afshar Sedigh et al. [1] this study has the following the following assumptions:

1. Constant delivery time;

2. Poisson process with a constant, known, and identical rate of customer demand arrival;

3. Backlogged shortage;

4. The availability of online information about the retailer's inventory position and demand for the central warehouse.

We wish to state some reasons behind these assumptions. First, it is worth noting that Poisson arrival rate and constant delivery time extensively employed in literature; for instance, Axsäter [2, 3, 4], Axsäter & Marklund [5], and Forsberg [15, 16] are studies that considered both constant delivery time and Poisson/compound Poisson arrival rates. Also, several studies addressed retailers with identical demand rates to facilitate their modelling (e.g. [39, 7, 29, 13]).

Backlogged shortage is a popular policy (e.g. see [32, 32, 18]). Also, a known batch size along with a negligible ordering cost is widely used and acceptable in both literature and practice (e.g. see [2, 3, 4, 5, 15, 16, 19, 38, 12, 28]). The reasons for such an approach include packaging or shipping requirements because of economies of scale in handling or shipping limitations, along with negligible shipping costs or benefiting from electronic commerce.

Finally, there are instances indicating that central warehouse had access to the information about retailers' demand activities and inventory status for more than two decades – Kurt Salmon Associates

[22, 23] reported employment of information technology in grocery industries, or Stalk et al. [37] attributes Wal-Mart success to reasons such as detailed information sharing of customer's behavior. Therefore, it is acceptable to assume that the central warehouse has access to information about branches' inventory on hand, especially due to the availability of high-speed information exchange infrastructures and automated stores.

Having said the justifications, we restate Moinzadeh's [28] policy; retailers use the (R, Q) ordering policy, the central warehouse starts with m initial batches ($m \geq 0$) and adopts the following ordering policy: Immediately after a retailer's inventory position reaches $R+s$, ($0 \leq s \leq Q-1$), a batch from an external supplier is ordered.

3 An Illustrative Example

Here we restate the example for two retailers from our earlier study (i.e. [1, Section 3]) to obtain the central warehouse's inventory position. As can be seen in Figure 3, immediately after a retailer's inventory position reaches $R+s$, the central warehouse's inventory position is whether $(m + 1)Q$ or $(m + 2)Q$, considering inventory position of retailers at $t-$ (Figure 7(A) and (B)).

For the rest of this section let $m = 3, Q = 4$, and $s = 2$; Figure 4(B) exemplifies the case that the retailer who initiates ordering Q_0 and the retailer who orders Q_0 are different. Overall, the retailers who initiates ordering Q_0 (i.e. the batch we are interested in) and the retailer to whom the batch is shipped are independent. For example, consider the situation that the inventory position of the retailer 1 is $R+s(R+2)$ at t_0 (Figure 4(A) and (B)) – i.e. it initiates ordering Q_0. However, Q_0 can be shipped to Retailer 1 (Figure 4(A)) or Retailer 2 (Figure 4(C)). Henceforth, we address the cases that inventory position of Retailer 1 equals $R + 2$ at t_0.

Let Retailer 1 order Q_0, an illustration of this system is presented in Figure 5. We know that between t_0 and t_1^- three batches are ordered (and a batch at t_1), and Retailer 1's inventory position equals $R + 1$ at t_1^-. Knowing that Retailer i can order b_i batches, so that

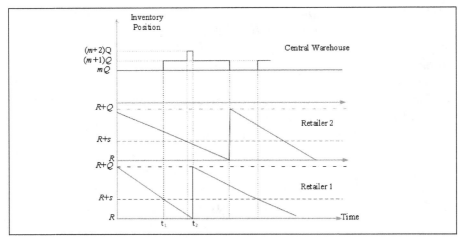

Figure 3: The central warehouse inventory position

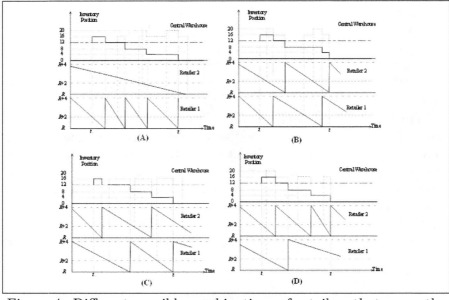

Figure 4: Different possible combinations of retailers that cause the central warehouse to demand Q_0 and the retailer that demands Q_0 from the central warehouse when $I_w(t_0^-) = mQ$

$b_1 + b_2 = 3, b_i \geq 0, \forall i$, we want to calculate the probability of ordering these three batches by 1 demands. Besides we know that, $l = l_1 + l_2$, and l_r demands cause the retailer r to order b_r batches from the central warehouse. Let $l_r = b_r Q + u_r$, we need to obtain the probability that retailer r receives l_r demands of total l demands during $t_1^- - t_0$ and the probability that these l_r demands initiate b_r batches ordering, i.e. $l_r = b_r Q + u_r$.

The probability of distributing l demands between two retailers as $l = l_1 + l_2$ equals to $(l/l_1)(1/2)^l$ for identical demand rates. Moreover, the last demand should occur at Retailer 1. Thus, we eliminate the last demand. To do so, let us assume $l' = l + 1$, $[l']_1 = l_1 + 1$ and $l' = [l']_1 + l_2$ and the probability is calculated as follows:

$$1/2(l/l_1)(1/2)^l = 1/2((l'-1)/([l']_1 - 1))(1/2)^{(l'-1)} = \qquad (1)$$
$$((l'-1)/([l']_1 - 1))(1/2)^{(l')}$$

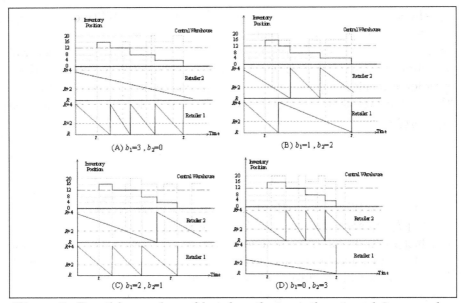

Figure 5: Possible number of batches that retailers 1 and 2 can order when Retailer 1 causes the central warehouse to order Q_0 and consumes this batch itself

Let $R + k'$ show the inventory position of Retailer 2 at t_0. In the example in question, we can calculate the upper and lower bounds of l'. To obtain the lower bound, inventory position of Retailer 2 at $[t_1]^-$ equals to $R + Q$; therefore

$$l' \geq k' + 4(b_2 - 1) + 4b_1 + 2 \qquad (2)$$

We know that $k' \geq 3$ and $b_1 + b_2 = 3$; therefore, $l' \geq 13$. Note that in Figure 5(A) the lower bound cannot be met, since $b_1 = 3$ and the lower bound is $4b_1 + 2 = 14$, i.e. $l' = 13$ is not feasible.

To obtain the upper bound, we note that Retailer 2's inventory position should be equal to $R + 1$ at $[t_1]^-$, and we have:

$$l' \leq k' - 1 + 4b_2 + 4b_1 + 2 \qquad (3)$$

On the other hand, we know that $k' \leq 4$ and $b_1 + b_2 = 3$; therefore, $l' \leq 17$.

Finally, $l' = [l']_1 + l_2$ does not guarantee that all the conditions to be held. For example, let $[l']_1 = 7$ and $l_2 = 9$; although $13 \le l' = 16 \le 17$, the inventory position of Retailer 1 at $[t_1]^-$ equals $R + 4$. On the other hand, let $[l']_1 = 14$ and $l_2 = 3$, the inventory position of Retailer 1 at $[t_1]^-$ equals $R + 1$; but with probability $1/2, k' = 4$ which leads to a feasible solution. Therefore, we need to multiply (1) by some other probability functions to obtain the probability of demanding 3 batches by these l' demands. In other words, for retailer r we should calculate the probability of ordering b_r batches by l_r demand (note that $l_r = b_r Q + u_r$); we show this as $P(b_r Q + u_r \to b_r)$. Two illustrative computations for $P(b_2 Q + u_2 \to b_2)$ are presented in Figure 6. In Figure 6, we show the case that Retailer 2's inventory position is more than $R + 2$ at t_0 (i.e. it is either $R + 3$ or $R + 4$). We show the two different cases with distinctive lines. Figure 6(A) shows the calculation of $P(b_2 Q + u_2 \to b_2)$ for $u_2 = -1 (Q = 4$ and $b_2 = 3)$, i.e. $P(11 \to 3)$. As can be seen, only one of the two cases is feasible the other case leads to ordering two batches; hence, the probability is $1/2$. Figure 6(B) shows the calculation of $P(b_2 Q + u_2 \to b_2)$ for $u_2 = 1$ and the same Q and b_2, i.e. $P(13 \to 3)$. Here both cases are feasible; hence, the probability is 1.

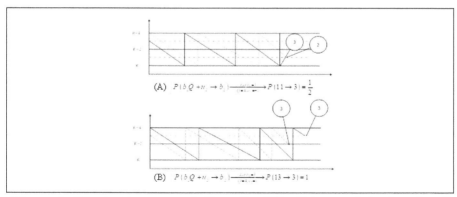

Figure 6: Illustrative computations for $P(b_2 Q + u_2 \to b_2)$ when $b_2 = 3$ and $u_2 = -1$ or $u_2 = 1$

4   Mathematical Model

This model extends Afshar Sedigh et al. [1] for more than two retailers and obtains cost function of Moinzadeh [28]. In this section, we obtain the probability distribution and boundaries for customer demand. Finally, we obtain the mathematical model for this problem.

4.1   Probability distribution of customer demand (k)

The first step to obtain the probability distribution of the customer demand is to find the inventory positions of the central warehouse and the retailers at t_0^-. Note that at this moment (i.e. t_0^-) one of the retailer's inventory positions is $R+s$ (see t_2 in Figure 3). Furthermore, the inventory position at the central warehouse at this moment is a function of Q, m, and $|A_0(t_0^-)|$, where $|A|$ is used to indicate the cardinality (number of members) of set A. In what follows, we obtain the inventory position at the central.

4.1.1   The inventory position of central warehouse

We wish to remind the readers that $A_0(t_0^-)$ is the set of retailers whose inventory positions at t_0^- are less than or equal to $R+s$. Also, we know that inventory positions of retailers are uniformly distributed between $R+1$ and $R+Q$ [17]. We say the inventory system operates in state i if $|A_0(t_0^-)| + 1 = i$. In order to obtain the probabilities associated with the system states, let $I_a(t_0^-)$ be the inventory position of retailer a at t_0^-. Note that:

$$P(a \in A_0(t_0^-)) = P(R+1 \le I_a(t_0^-) \le R+s) = s/Q$$
$$P(a \in A_1(t_0^-)) = P(R+s+1 \le I_a(t_0^-) \le R+Q) = (Q-s)/Q \quad (4)$$

Thus, the probability that the system operates in state i at t_0, i.e. $p(i, s)$, is:

$$P(i, s) = (N - 1i - 1)(s/Q)^{(i-1)}((Q - s)/Q)^{(N-i)}; s > 0 \quad (5)$$

In state i, when $s > 0$, the inventory position of the central warehouse at t_0^+ is $(m+i)Q$. Furthermore, because the retailers are assumed to be identical, without losing generality the following sequence is assumed for the retailers (let a be a retailer's number):

$$a \in A_0(t_0^-) \text{ if } a \leq i - 1 I_a(t_0^-) = R + s \text{ if } a = i a \in A_1(t_0^-) \text{ if } a \geq i + 1 \tag{6}$$

In the above mentioned equation, for a system entering state i at t_0, there are i retailers with an inventory position with less than or equal to $R + s$. From the stated retailers one surely has an inventory position equal to $R + s$ (the one whom triggered the order by the central warehouse); we call it retailer i. For the rest of retailers, if their inventory position is less than or equal to $R + s$, they have a number from 1 to $i - 1$, else $i + 1$ to N. Furthermore, we wish to remind the reader that we call the retailer who orders Q_0 and the time of ordering Q_0, B and t_1, respectively. The inventory position of retailer B can take one of the following values based on occurrence of three mutually exclusive events:

$$
\begin{aligned}
& [R + 1 \leq I]_B(t_0) < \\
& \qquad R + s \, if \, B \leq i - 1 I_B(t_0) = R + s \\
if B = i & [R + s + 1 \leq I]_B(t_0) \leq R + Q \\
& \qquad if B \geq i + 1
\end{aligned}
\tag{7}
$$

Let the system be in state i, where there are $m + i$ batches in the central warehouse at t_0^+. We know that the last customer demand for Q_0 had occurred at t_1^- and the inventory position of B decreased to R at t_1. Now we are ready to calculate the probability that b batches are ordered by retailer r. The following two notations are used for this purpose:

$\mu_{l_B}^{ir}$: A random variable that represents the number of batches ordered by retailer r when the system is in state i and retailer B orders Q_0 to fulfil l demands

$\eta_{l_B}^{ir}$: A random variable that shows the number of batches ordered by the first r retailers when the system is in state i and retailer B orders Q_0 after arriving l demands to all retailers.

Note that the last customer demand is disregarded (it is assigned to retailer B). In this case, the schema of the inventory system is modified as shown in Figure 7. Therefore, the following recursive relation is used to determine the probabilities associated with different values of $\eta_{l_B}^{ir}$

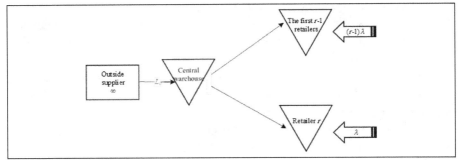

Figure 7: System with reduced retailers

Because inventory position of each retailer is renewed every Q customer demand (see Figure 3), evaluating system demands will be easier using $bQ + u$ instead of l (i.e. we calculate $Pr(\mu_r^i, bQ + u, B = b)$).

This probability depends on s, residual inventory (u), system state (i), and retailer (r). When the value of parameters stated earlier are given, the inventory position of retailer B is defined in one of the intervals defined in (7).

4.1.2 Calculating the probabilities for retailer who orders Q zero

Here we study when $s > 0$ and $r = B$, retailer B orders b batches, when $bQ + u + I_B$ (t_0) equals $bQ + R + 1$ (the last demand should lead to ordering the last batch).

Then using relation (7), it is possible to obtain acceptable values of u, based on which $Pr(\mu_{r,bQ+u,B}^i)$ is obtained by:

$$Pr(\mu^i_{r,bQ+u,B} = b) = \{\frac{1}{s}B < i, s > 0, r = B$$

$$0 \leq u \leq s - 1, b \geq 0, 1,$$

$$B = i, s > 0, r = B, u = s - 1, b \geq 0, \frac{1}{(Q-s)}, B > i, s > 0, r = B \tag{8}$$

$$s \leq u < Q, b \geq 0\}$$

4.1.3 Calculating the probabilities for retailer who does not order Q_0

When $s > 0$ and $r \neq B$, u can take a value in one of the two intervals $0 \leq u \leq I_r(t_0) - R - 1$ and $I_r(t_0) - R \leq Q + u < Q$.

Therefore, we have:

$$I_r(t_0) - Q - R \leq u \leq I_r(t_0) - 1 - R. \tag{9}$$

Then, using relations (7) and (11), the acceptable intervals of u is obtained as follows:

$$\{-Q + 1 \leq u \leq s - 1 \text{ if } r \leq i - 1, 1 \leq u \leq s \text{ if}$$
$$r = i \ s + 1 - Q \leq I_B(t_0) \leq Q - 1 \text{ if } r \geq i + 1\} \tag{10}$$

Relation (12) states that $-Q + 1 \leq u \leq s - 1$ for s>0, $r \neq B$, and $r \leq i - 1$. It means that u can take a value either in (I) $0 \leq u \leq s - 1$ or in (II) $-Q + 1 \leq u \leq -1$. The former occurs if $I_r(t_0) > R + u($ u should be less than $I_r(t_0) - R)$. As $I_r(t_0)$ is uniformly distributed in R+1,...,R+s, this event occurs with probability of:

$$Pr(\mu^i_{r,bQ+u,B} = b) = \frac{(s - u)}{s}; B < i, s > 0, r \neq B, 0 \leq u \leq s, b \geq 0 \tag{11}$$

In order to obtain $Pr(\mu^i_{r,bQ+u,B} = b)$, for the interval defined in (II), two cases are considered as stated in the following.

(II-a) $s - Q < u \leq -1$; here the customer demand is $u + Q$; hence, the interval can be modified as $s < u + Q \leq Q$. This demand surely causes a retailer to order, because it exceeds $I_r(t_0)$. Thus, the corresponding probability is 1. In other words, we have:

$$Pr(\mu^i_{r,bQ+u,B} = b) = 1; B < i, s > 0, r \neq B, s-Q < u < 0, b \geq 1 \quad (12)$$

$(II - b) - Q + 1 < u \leq s - Q$; in this situation $0 < u + Q \leq s$ and hence the customer demand causes a retailer to order if it is more than or equal to $I_r(t_0) - R$. Thus, the probability that $u + Q \geq I_r(t_0)$ is obtained as follows:

$$Pr(\mu^i_{r,bQ+u,B} = b) = \quad (13)$$
$$(Q + u)/s; B < i, s > 0, r \neq B, -Q < u \neq s - Q, b > 0$$

Similarly, the following probabilities are obtained:

$$Pr(\mu^i_{(r}, bQ + u, B)^i = b) = 1 + u/(Q - s); r > i,$$
$$r \neq B, s > 0, s - Q < u \leq 0, b \geq 1$$
$$r > i, r \neq B, s > 0, s - Q < u \leq 0, b \geq 0 (Q - u)/(Q - s); \quad (14)$$
$$r > i, r \neq B, s - Q < u \leq 0,$$
$$b \geq 0, r = i, r \neq B, s > 0, s - Q < u \leq 0, b \geq 1; r = i,$$
$$r \neq B, s > 0, s - Q < u \leq 0, b \geq 0$$

4.1.4 The inventory position of central warehouse at t_0^1 when $s = 0$

When $s = 0$, the system surely operates in state 1 at t_0. In other words, without information sharing, the probability that the system operates in state i at t_0 is obtained as:

$$p(i, s) = \{(N-1i-1)(s/Q)^{(i-1)}((Q-s)/Q)^{(N-i)}s > 01i = 1, s = 0\} \quad (15)$$

Note that $Pr(\mu_{(r}, bQ + u, B)^i = b)$ is meaningful when $B \geq 1$. Otherwise, the last demand must be considered in order to avoid negative customer demand, when $B = 1$ and $b = 0$. It is also obvious that at t_0, one batch is ordered by retailer 1 and the central warehouse, simultaneously. Thus, at t_0^+ the inventory position of the central warehouse is m and that $I_1(t_0^+) = Q$.

The probability of demanding b batches by retailer r is as follows:

$$
\begin{aligned}
Pr(\mu(r, bQ + u, B)^i = b) &= 1 \\
B = 1, s = 0, r = B, u = Q - 1, b &\geq 0, 1/Q \\
B > 1, s = 0, r = B, 0 \leq u < Q, b &\geq 0 \\
(Q + u)/Q; r > 1, r \neq B, s = 0, -Q < u \leq 1, b &\geq 1(Q - u)/Q; \\
r > 1, r \neq B, s = 0, 0 \leq u, Q, b &\geq 01; \\
r = 1, r \neq B, s = 0, 0 \leq u and Q, b &\geq 0
\end{aligned} \tag{16}
$$

Let the system be in state i, $m'(i, s)$ is defined as:

$$
m'(i, s) = m + is > 0 \tag{17}
$$

therefore, we need to compute the probability of ordering m' batches to fulfil exactly k customer demands. As mentioned in computing $Pr(\mu_(r, bQ + u, B, s)^i = b)$, we did not consider one customer demand to ensure that this customer demand was the last one. In our model, this demand will be assigned to retailer B, with probability of $1/N$. Therefore, the probability is calculated as:

$$
\tfrac{1}{N} Pr(\eta^i_{(N,l,B)} = m'(i, s) - 1)
$$

4.2 Boundaries of Customer Demand

Let us use m' instead of $m'(i, s)$. Thus, the following lemma is proposed:

Lemma 1. When the system has N retailers, from m' batches at least $m' - N$ batches are ordered by demand sizes equal to Q.

Proof. This lemma states that at most N batches can be ordered by

189

less than Q customer demand. Note that, when a retailer demands a batch, it meets its reorder point (R). Furthermore, each retailer at most meets R once with less than Q customers order, and after that its inventory position equals $R + Q$. Consequently, the number of retailers (N) gives the maximum number of the ordered batches with less than Q customer demand.

Now let us introduce the following notations:

$lb(i, s, m)$: The lower bound of customer demand, when system starts by m initial batches and is in state i for a given s.

$ub(i, s, m)$: The upper bound of customer demand, when system starts by m initial batches and is in state i for a given s.

Consider five subsystems as shown in Figure 5. This is based on differences in system behavior when $s > 0$ or $s = 0$ and the aforementioned lemma. The subsystems are stated in the following:

(A) $s > 0, m + I \geq N, i = N$;

(B) $s > 0, m + I \geq N, i < N$;

(C) $s > 0, m + I < N$;

(D) $s = 0, m < N$;

(E) $s = 0, m \geq N$.

4.2.1 Lower Bounds

When $s > 0$ and the system is in state i, for all subsystems we know that:

$$\min I_r(t_0) = R + 1r \leq i - 1R + sr = iR + s + 1r \geq i + 1.$$

Here $i - 1$ batches can be ordered by one customer demand to each one of the first $i - 1$ retailers. The other batches are divided into $m_1 = \min(N - i + 1, m)$ and $m_2 = \max(m - N + i, 0)$. First batch from the m_1 batches is ordered by s customer demand to retailer i

190

and remained batches are ordered by $s + 1$ customer demand to the rest of $N - i$ retailers.

It is obvious that the rest of m_2 batches are ordered by $m_2 Q$ customer demand to retailers. Therefore, the lower bounds when $s > 0$ is obtained as:

$$lb(i, s, m) = i - 1 + s + m(s + 1) \tag{18}$$

$$m+i < N, s > 0i-1+s+(N-i)(s+1)+(m+i-N)Qm+i \geq N, s > 0 \tag{19}$$

When $s = 0$, at most $N - 1$ batches are ordered by one customer demand and remained batches, if there is any, are ordered by Q customer demand. Therefore, the lower bounds are obtained as follows:

$$lb(i, s, m) = m < N, s = 0N - 1 + (m - N + 1)Qm \geq N,) \tag{20}$$
$$s = 0i - 1 + s + m(s + 1m + i < N,)$$
$$s > 0i - 1 + s + (N - i)(s + 1) + (m + i - N)Qm + i \geq N, s > 0$$

4.2.2 Upper Bound

When $s > 0$ and system is in state i, for all subsystems we know:

$$ub(i, s, m) = maxI_r(t_0) = R + sr \leq i - 1I_r$$
$$(t_0) = R + sr = imaxI_r(t_0) = R + Qr \geq i + 1 \tag{21}$$

To calculate the upper bound, it is assumed that $I_r(t_0)s$ are at their maximum level for all retailers ($\forall r$). Also, we divide the system into $i < N$ and $i = N$. In the first case, there always is a retailer with $I_r(t_0) > R + s$. Therefore, all the $m + i$ batches can be ordered by one of the $N - i$ last retailers with $(m + i)Q$ demand. Moreover, we assume that the rest of retailers have max $I_r(t_0) - R - 1$ demand. For the second case, $i = N$, whereas $[\max]_{(\forall i)} I_r(t_0) = s$, one of the batches is ordered by s demand. The rest is the same as previous case. Finally, the only difference between $s = 0$ and $s > 0$ is in batches quantities. As stated earlier, when $s = 0$, there is m batches in the central warehouse instead of $m + 1$.

$$ub(i, s, m) = i(s - 1) + (N - i - 1)(Q - 1)+ \\ (m + i)Qs > 0, i < N \\ (N - 1)(s - 1) + s + (m + i - 1)Qs > 0, \\ i = N(N - 1)(Q - 1) + mQ, s = 0 \tag{22}$$

4.2.3   Final Model

Before representing the final model, note that we consider an arbitrary unit of Q_0, let it be j^{th} unit of Q_0, with probability of $1/Q$. Furthermore, customer demand needed to order this unit between t_0^+ and t_1 equals $R + j$.

For computational simplicity, the following notation is introduced:

$f(i, B)$: Number of retailers which are in the same situation as $r = B$, when system is in state i.

Because we have identical retailers, the problem is simplified by evaluating system costs, based on which retailer $i - 1$, i or $i + 1$ orders Q_0 and then we multiply it by $f(i, B)$.

$$f(i, B) = i - 1B = i - 11B = iN - iB = i + 1 \tag{23}$$

Given the aforementioned, the final model is obtained as follows:

5   Conclusion and discussion

In this paper, we obtained the exact model for a two-echelon inventory system with a central warehouse and a number of retailers with Poisson distributed demand which was an open question. In the model unfulfilled customer demand was backlogged. The batch size was given, for retailers and the central warehouse. Using conditional probabilities, we could derive the exact model for identical retailers. The model facilitates the optimization of Moinzadeh's [28] policy and extending its model to more general scenarios (e.g. retailers with different delivery time).

Moinzadeh's [28] policy has the benefit of simplicity for implementation that makes it a good candidate to utilize in the industries that does not want to invest in a complex policy such as the one proposed by Axsäter & Marklund [5].

This model can also be extended for retailers with unidentical demand rates; however, the extension would be more complex. Finally, it is worth noting that, although using a simple to implement policy may not reduce the costs the same as more flexible policies, such a policy enables more medium-sized businesses to utilize information sharing and reduce their costs in comparison with otherwise not sharing information in the supply chain.

References

[1] Afshar Sedigh, A. H., Haji, R., & Sajadifar, S. M. (2019). Cost function and optimal boundaries for a two-level inventory system with information sharing and two identical retailers. Scientia Iranica. Transaction E, Industrial Engineering, 26(1), 472-485.

[2] Axsäter, S. (1990). Simple solution procedures for a class of two-echelon inventory problems. Operations research, 38(1), 64-69.

[3] Axsäter, S. (1993). Exact and approximate evaluation of batch-ordering policies for two-level inventory systems. Operations research, 41(4), 777-785.

[4] Axsäter, S. (2000). Exact analysis of continuous review (R, Q) policies in two-echelon inventory systems with compound Poisson demand. Operations research, 48(5), 686-696.

[5] Axsäter, S., & Marklund, J. (2008). Optimal position-based warehouse ordering in divergent two-echelon inventory systems. Operations Research, 56(4), 976-991.

[6] Axsäter, S., & Rosling, K. (1993). Installation vs. echelon stock policies for multilevel inventory control. Management Science, 39(10), 1274-1280.

[7] Bradley, J. R. (2017). The Effect of Distribution Processes on Replenishment Lead Time and Inventory. Production and Operations Management, 26(12), 2287-2304.

[8] Chen, F., & Zheng, Y. S. (1994). Evaluating echelon stock (R, nQ) policies in serial production/inventory systems with stochastic demand. Management Science, 40(10), 1262-1275.

[9] Chen, F., & Zheng, Y. S. (1998). Near-optimal echelon-stock (R, nQ) policies in multistage serial systems. Operations research, 46(4), 592-602.

[10] Chen, M. C., Yang, T., & Yen, C. T. (2007). Investigating the value of information sharing in multi-echelon supply chains. Quality & quantity, 41(3), 497-511.

[11] Cho, D. W., & Lee, Y. H. (2013). The value of information sharing in a supply chain with a seasonal demand process. Computers & Industrial Engineering, 65(1), 97-108.

[12] Deuermeyer, B. L., & Schwarz, L. B. (1979). A model for the analysis of system service level in warehouse-retailer distribution systems: the identical retailer case. Institute for Research in the Behavioral, Economic, and Management Sciences, Krannert Graduate School of Management, Purdue University.

[13] Fleischmann, M., Kloos, K., Nouri, M., & Pibernik, R. (2020). Single-period stochastic demand fulfillment in customer hierarchies. European Journal of Operational Research.

[14] Forrester, J. W. (1958). Industrial Dynamics. A major breakthrough for decision makers. Harvard business review, 36(4), 37-66.

[15] Forsberg, R. (1995). Optimization of order-up-to-S policies for two-level inventory systems with compound Poisson demand. European Journal of Operational Research, 81(1), 143-153.

[16] Forsberg, R. (1997). Exact evaluation of (R, Q)-policies for two-level inventory systems with Poisson demand. European journal of operational research, 96(1), 130-138.

[17] Hadley, G., & Whitin, T. M. (1963). Analysis of inventory systems. Englewood Cliffs, N.J: Prentice-Hall.

[18] Halim, M. A. (2017). Weibull distributed deteriorating inventory model with ramp type demand and fully backlogged shortage. Asian Journal of Mathematics and Computer Research, 148-157.

[19] Haji, R., & Haji, A. (2007). One-for-one period policy and its optimal solution. Journal of Industrial and Systems Engineering, 1(2), 200-217.

[20] Haji, R., & Sajadifar, S. M. (2008). Deriving the exact cost function for a two-level inventory system with information sharing. Journal of Industrial and Systems Engineering, 2(1), 41-50.

[21] Janaki, D., Izadbakhsh, H., & Hatefi, S. (2018). The evaluation of supply chain performance in the Oil Products Distribution Company, using information technology indicators and fuzzy TOPSIS technique. Management Science Letters, 8(8), 835-848.

[22] Kurt Salmon Associates. (1993). Efficient Consumer Response:[ECR]; enhancing consumer value in the grocery industry. Research Department Food Marketing Institute, Kurt Salmon Associates, Inc., Washington, Dc (1993).

[23] Kurt Salmon Associates. (1997). Quick Response: Meeting Customer Need. Kurt Salmon Associates, Atlanta, GA.

[24] Li, M., & Simchi-Levi, D. (2020). The Web Based Beer Game-Demonstrating the Value of Integrated Supply-Chain Management. In MIT forum for Supply Chain Management Web site. Retrieved July.

[25] Li, X., & Wang, Q. (2007). Coordination mechanisms of supply chain systems. European journal of operational research, 179(1), 1-16.

[26] Liu, C., Huo, B., Liu, S., & Zhao, X. (2015). Effect of information sharing and process coordination on logistics outsourcing. Industrial Management & Data Systems.

[27] Liu, C., Xiang, X., & Zheng, L. (2020). Value of information sharing in a multiple producers–distributor supply chain. Annals of Operations Research, 285(1-2), 121-148.

[28] Moinzadeh, K. (2002). A multi-echelon inventory system with information exchange. Management science, 48(3), 414-426.

[29] Nakade, K., & Yokozawa, S. (2016). Optimization of two-stage production/inventory systems under order base stock policy with advance demand information. Journal of Industrial Engineering International, 12(4), 437-458.

[30] Ojha, D., Sahin, F., Shockley, J., & Sridharan, S. V. (2019). Is there a performance tradeoff in managing order fulfillment and the bullwhip effect in supply chains? The role of information sharing and information type. International Journal of Production Economics, 208, 529-543.

[31] Sajadifar, S. M., & Haji, R. (2007). Optimal solution for a two-level inventory system with information exchange leading to a more computationally efficient search. Applied mathematics and computation, 189(2), 1341-1349.

[32] San-José, L. A., Sicilia, J., & Alcaide-López-de-Pablo, D. (2018). An inventory system with demand dependent on both time and price assuming backlogged shortages. European Journal of Operational Research,

270(3), 889-897.

[33] San-José, L. A., Sicilia, J., González-De-la-Rosa, M., & Febles-Acosta, J. (2019). Analysis of an inventory system with discrete scheduling period, time-dependent demand and backlogged shortages. Computers & Operations Research, 109, 200-208.

[34] Simchi-Levi, D., Kaminsky, P., Simchi-Levi, E., & Shankar, R. (2008). Designing and managing the supply chain: concepts, strategies and case studies. Tata McGraw-Hill Education.

[35] Simchi-Levi, D., & Zhao, Y. (2007). Three generic methods for evaluating stochastic multi-echelon inventory systems. Working paper, Massachusetts Institute of Technology, Cambridge.

[36] Simchi-Levi, D., & Zhao, Y. (2012). Performance evaluation of stochastic multi-echelon inventory systems: A survey. Advances in Operations Research, 2012.

[37] Stalk, G., Evans, P., & Shulman, L. E. (1992). Competing on capabilities: The new rules of corporate strategy. Harvard business review, 70(2), 57-69.

[38] Svoronos, A., & Zipkin, P. (1988). Estimating the performance of multilevel inventory systems. Operations Research, 36(1), 57-72.

[39] Wang, G., & Gunasekaran, A. (2017). Operations scheduling in reverse supply chains: Identical demand and delivery deadlines. International Journal of Production Economics, 183, 375-381.

[40] Zhou, H., & Benton Jr, W. C. (2007). Supply chain practice and information sharing. Journal of Operations management, 25(6), 1348-1365.

Appsent: A Tool That Analyzes App Reviews

Chan Won Lee

University of Otago, Dunedin, New Zealand

chacha04@hotmail.com

Saurabh Malgaonkar

University of Otago, Dunedin, New Zealand

malsa876@student.otago.ac.nz

Sherlock A. Licorish

University of Otago, Dunedin, New Zealand

sherlock.licorish@otago.ac.nz

Bastin Tony Roy Savarimuthu

University of Otago, Dunedin, New Zealand

tony.savarimuthu@otago.ac.nz

Amjed Tahir

School of Fundamental Sciences, Massey University, New Zealand

a.tahir@massey.ac.nz

Abstract

Enterprises are always on the lookout for tools that analyze end-users' perspectives on their products. In particular, app reviews have been assessed as useful for guiding improvement efforts and software evolution, however, developers find reading app reviews to be a labor intensive exercise. If such a barrier is eliminated, however, evidence shows that responding to reviews enhances end-users' satisfaction and contributes towards the success of products. In this paper, we present Appsent, a mobile analytics tool (as an app), to facilitate the analysis of app reviews. This development was led by a literature review on the problem and subsequent evaluation of current available solutions to this challenge. Our investigation found that there was scope to extend currently available tools that analyze app reviews. These gaps thus informed the design and development of Appsent. We subsequently performed an empirical evaluation to validate Appsent's usability and the helpfulness of analytics features from users' perspective. Outcomes of this evaluation reveal that Appsent provides user-friendly interfaces, helpful functionalities and meaningful analytics. Appsent extracts and visualizes important perceptions from end-users' feedback, identifying insights into end-users' opinions about various aspects of software features. Although Appsent was developed as a prototype for analyzing app reviews, this tool may be of utility for analyzing product reviews more generally.

keywords: requirements engineering; software maintenance; customer feedback; customer experience; data analytics; natural language processing; sentiment analysis; multidimensional analysis; emotion analysis; app-reviews

1 Introduction

The mobile app market was estimated to be a 77 billion dollar industry in 2017, with more than five million apps hosted on Online Application Distribution Platforms (OADPs) [1, 2]. This is linked to massive sales of mobile devices and popularity of their usage worldwide [3]. The commonly accessed OADPs are Google play store, Apple app store and Windows apps [4-6]. OADPs allow developers to host their apps for mobile device users, facilitating app updates when developers release new versions of their apps. Another useful feature provided by OADPs is the ability to support direct communication between app users and developers through reviews. These reviews contain important feedback related to apps' performance from users' perspectives.

However, OADPs host numerous reviews, which are open to public access in informing future users' decisions in relation to app use. Beyond a star rating, reviews normally contain complaints about commonly faced problems, users' sentiments about features, suggestions for improvement, and requests for new features [4]. Thus, in meeting the expectations of users, app developers must analyze users' reviews to evolve their apps. This knowledge also significantly assists developers in their user-driven software quality evaluation and product marketing process [5]. However, manually processing large volumes of reviews demands high levels of cognitive load and time if it is performed manually [6, 7]. In fact, this burden may also be compounded due to ambiguity and sarcasm present in the reviews [8, 9]. Thus, a combination of natural language processing and data mining techniques have been recommended for addressing such challenges [9-11]. Such techniques are held to have promise in terms of assisting developers with extracting and visualizing meaningful information from user reviews [12]. This way, developers may quickly identify issues faced by users and discern the source of their (dis)satisfaction, in enhancing the quality of their app and gaining a competitive edge in the app market [13, 14]. We have looked to validate this opportunity in this work, and developed Appsent to provide these details for developers. We introduce Appsent in this paper, whose primary objective is to provide instant real-time meaningful interpretation of the information present in the app reviews. Although being a prototype, outcomes of Appsent's evaluations suggest that this tool could be of utility to app developers. We present our portfolio of work around the development of Appsent and its evaluation, contributing practical insights for developers, and understandings for the software engineering community in terms of how tools may be engineered to rapidly support the evolution of software. The remaining sections of this paper are organized as follows. In Section II, we review relevant literature and tools for analyzing user reviews. This leads to the identification of research gaps and suitable research questions. Section III describes the design and implementation aspects of Appsent. Section IV presents our empirical evaluation of Appsent, followed by the tool's overview and evaluation outcomes in Section V. We discuss our findings and implications of the work in Section VI, before considering threats to the work in Section VII. Finally, we provide concluding remarks in Section VIII.

2 Background

Text mining as a discipline is held to offer an efficient solution to the problem of processing large volumes of textual information to extract knowledge [15]. The text mining process is data-driven, and assists significantly in identifying the hidden patterns or trends in unstructured text [16]. Central to the meaningful interpretation of textual information is the ability to identify the polarity of users. The primary objective of polarity analysis is to discover the attitude of the user towards specific things or events [17]. Polarity can be evaluated via sentiment (e.g., positive or negative) or emotion (e.g., happy or fear). The polarity of a sample of text may be uncovered by using learning- or lexicon-based methods [18]. Learning methods build a feature classification model by utilizing machine learning techniques or probabilistic models, whereas lexicon methods use well-built dictionaries to determine the meaning of textual entities. Another critical stage of text mining is topic or feature identification and extraction. Part-of-speech (POS) tagging and n-gram analysis are commonly used for extracting features or topics of interest in textual data [19]. For instance, Licorish et al. [6] employed these approaches to identify the most frequently mentioned features in enhancement requests logged by the Android community. In another study, Iacob et al. [19] were successful in discovering associations among features through the use of these approaches and Latent Dirichlet Allocation (LDA) models. A more advanced study by Lee et al. [20] has deployed POS tagging, n-gram analysis and Social Network Analysis (SNA) techniques to identify features and explore associations among features that were mentioned in reviews. This latter study places emphasis on the importance of discovering the relations between features in understanding the complexities of defects and improvement pointers that are identified by users in reviews. The abovementioned studies automatically process reviews to provide a high-level view of the information of interest to users. Beyond understanding features that capture users' interests, such provisions enable developers to identify the complex relationships between app features, and significantly assists in co-relating users' problems. We anticipated that tools built to assist developers with extracting and visualizing meaningful information from user reviews should identify users' opinions about software features (e.g., positive views around specific features), identify the features of particular interest (e.g., broken features), and discover the associations among the features (e.g., which group of features affects each other). Browsing for such products on the app stores and wider internet it is observed that software enterprises have used the methods identified above to develop tools that fulfill the task of analyzing the performance of apps released on OADPs. After identifying four popular tools available online, we performed a systematic comparative study of these tools. The four tools are AppTrace, AppFigures, SensorTower, and Apptentive [21-24]. Table I highlights the systematic comparative study of the four tools. The tools have been classified based on their target audience, market pricing, domain of application, and a set of features. We compared the features organically, and also used dimensions suggested in [20] (noted below) for evaluating the tools utility for identifying users' opinions about software features, features of particular interest, and associations among features. We compared the following criteria:

- daily ranks: computed based on the number of daily downloads.

- country ranks: computed based on the number of downloads across countries.

- sentiment analysis and sentiment analysis with keywords: computed based on the polarity of reviews and the polarity that keywords attracted.

- sales analysis: computed based on revenue generated by the app to date.

201

- competitor analysis: computed based on the ranking, sentiment, and sales features.

- star ratings: the rating of the app on a numerical scale of one to five.

- advanced text mining option: the ability of the tool to discover meaningful associations among the entities of interests present in app reviews.

Table I. Summary of commercially available tools to analyze reviews

		App Trace	App Figures	Sensor Tower	Apptentive
		1 = Feature Present, 0 = Feature Absent			
Application Name		App Trace	App Figures	Sensor Tower	Apptentive
Target Audience		Product owner, Developer	Product owner, Developer	Product owner, Developer	Product owner, Developer
Price		Free	$9/month	$79/month (Personal) $399/month (Business)	$299/month
Domain		Smart phone Apps	Smart phone Apps	Smart phone Apps	Smart phone Apps
FEATURES	Daily Ranks	1	1	1	1
	Country Ranks	1	1	1	1
	Sentiment Analysis	1	1	1	0
	Sentiment Analysis with Keywords	1	0	0	0
	Sales Analysis	0	1	1	0
	Competitor Analysis	0	0	1	0
	Star Rating	0	0	0	1
	Advanced Text Mining (Feature relationship analysis)	0	0	0	0

We also informally evaluated if the tool was provided as an interface for mobile devices, given their popularity [1, 2]. In Table I, '0' denotes the absence of a feature while '1' indicates the presence of a feature in the particular tool.

Evaluating the tools it was observed that only Apptentive was available for use on smartphone (mobile) devices (Table I). However, overall, Apptentive supported fewer features than the other tools. By running the tools, we found that the sentiment analysis feature of AppTrace did not return readable visualizations at times (i.e., too much dimensions were evident). AppFigures required manual intervention for analyzing reviews, and SensorTower could not fully distinguish the sentiment for some reviews, requiring manual intervention. Overall, it was observed that these tools are still being evolved and they do not provide mechanisms for identifying associations among features, which is assessed as critical for identifying complex relationships among defects [20].

Therefore, we believe that having a tool that analyzes user reviews will allow developers to quickly identify issues faced by users and discern the source of their (dis)satisfaction, in enhancing the quality of their app and gaining a competitive edge in the apps market. However, examining previous works and tools available for analyzing reviews to inform developers, Table I shows that while tools available provide various functions to support reviews analysis, they lack various features that may be deemed pertinent to developers [20]. We thus proposed to provide a prototype app to bridge this gap. To guide this effort we formulated the following three research questions: **RQ1.** How can we design and develop a mobile app for analyzing reviews that is deemed highly usable? Upon the creation of the mobile app, we investigate the utility provided by the functionalities offered by the app. Thus, the second question we investigated is: **RQ2.** Does the newly developed app provide functionalities that are helpful? Any analytics tool that is developed should unearth insights from data and present it to the users. In this work, we investigate the ability of the tool to provide insights at different levels (e.g., based on sentiments and emotions expressed by users). Towards examining the nature of insights provided by the app, we pose the last question. **RQ3**. Does the developed app provide suitable interpretative analytics on multiple levels?

3   APPSENT - Design And Implementation

We provide details around the techniques and principles that were used to design and implement our app, Appsent, in this section towards answering RQ1.

3.1   Appsent Design

App reviews contain unstructured textual information that needs to be pre-processed for relevance [25]. This process involves the removal of blank spaces, unwanted characters, and stopwords. We performed this process, before aggregating similar words (i.e., stemming). Stopwords were identified through WordNet and Stanford NLP API was used for stemming based on Porter's algorithm [26, 27]. This process refined reviews making them suitable for further analysis. We next performed natural language processing through the use of POS tagging [28]. For this phase we used the Stanford API to extract the nouns and verbs in reviews; which map to features and issues respectively [29]. The extracted features and issues were tallied to quantify the frequency of each feature or issue. Sentiment analysis identifies the positive, negative or neutral tone of a statement made by users. To detect the sentiments in reviews we employed a sentiment analysis method that has been inspired by the sentiment treebank technique [30]. In exploring emotions in reviews we used WordNet's lexicon-based method to identify six emotions, namely happy, sad, surprise, fear, anger and disgust [31]. In checking the associations among features we performed co-occurrence analysis as proposed by Lee et al. [20]. This analysis captures the relationship between features and associated problems through feature-feature and feature-issue analyses. Appsent implements the same strategy and performs co-occurrence analysis by considering the sentiment, emotion, and the rating (1-5) factors.

After finalizing the functional aspects of Appsent, it was necessary to design a user-friendly graphical user interface (GUI). We considered both users' interaction with the app and the provision of meaningful visualizations of results on mobile device screens. In informing our design we adapted Shneiderman's rules [32]. These rules are provided in Table II, covering the creation of shortcuts of the most frequently visited features of the app to designing the app for enjoyment. Of particular interest to us was also the provision of informative responses, memory optimization, eliminating unwanted details, quick load of the application, and the provision of highly immersive and appealing interfaces (refer to Table II). Figure 1 shows the diagrammatic represen-

Table II. Appsent's User Interface Design Rules

ID.	Rule
1	Enable frequent users to use shortcuts
2	Offer informative feedback
3	Design dialogs to yield closure
4	Support internal locus of control
5	Consistency
6	Reversal of actions
7	Error prevention and simple error handling
8	Reduce short-term memory load
9	Design for multiple and dynamic contexts
10	Design for small devices
11	Design for limited and split attention
12	Design for speed and recovery
13	Design for "top-down" interaction
14	Allow for personalization
15	Design for enjoyment

tation of Appsent's system model, and highlights the various steps involved in the processing of app reviews. As illustrated in Figure 1, the unstructured textual data from reviews are pre-processed. This leads to a reduction in the size of data and removal of irrelevant data. After the data has been pre-processed, the data is then split into sentences, where each sentence is analyzed for emotional and sentiment states. Sentences present in reviews are parsed independently to identify the parts-of-speech. The POS tagger then marks the words (features or issues) of interests, and these words are stored in a database for further analysis. Thereafter, appropriate analysis mechanisms are executed to generate the information of interest. This information is stored in the smartphone's database. Appsent then queries the database as required, and the retrieved results are visualized using the appropriate data analytics charts. Thus, the smartphone app (Appsent) acts as a mobile carrier of the meaningful analytics information.

3.2 Appsent Implementation

Appsent was developed using Android studio and several open-source libraries [33]. To store data we used the SQLite database [34]. We employed two libraries for natural language processing; for POS tagging and sentiment extraction, we used Stanford CoreNLP 3.6.0 [35], and for performing emotion analysis we used Synesketch 2.0 [36]. For data visualization we used libraries from MPAndroidChart [37]. All the libraries used in this project are free for academic use. Appsent is built to operate on mobile devices running on Android version 3.2 (Honeycomb) or later, with a minimum of 1GB of memory.

4 APPSENT - Empirical Evaluation

In order to evaluate Appsent we engaged a company that designs mobile games in Dunedin, New Zealand called Runaway [38]. The company provided us with a dataset containing reviews taken from Google Play and Apple's App Store for one of their products. The product in question allows users to play a game involving nature. Our dataset contained 52,705 reviews with fifteen attributes associated with each review, including: Package Name, App Version Code, Reviewer Language, Device, Review Date, Star Rating, Review Title, and Review Text. We selected the Review Date, Star Rating, and Review Text at-

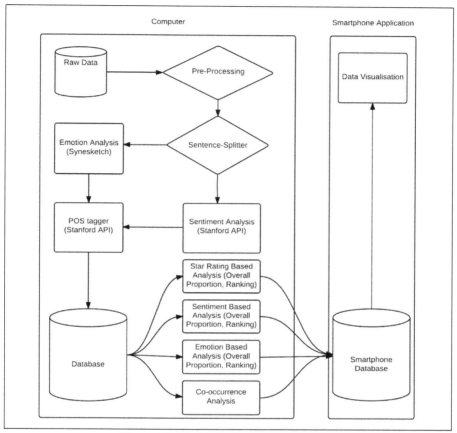

Figure 1: Appsent's system model

tributes for our evaluation of our prototype (Appsent). Review Date and Star Rating contained values that were well structured (e.g., "July 5, 2018" and "4"), and did not require any pre-processing. However, Review Text is mostly subjected to pre-preprocessing as mentioned in Section III(A). Beyond pre-processing, the processes highlighted in Figure 1 were followed towards visualizing analyzed reviews on Appsent's display screens.

To gain quality feedback on Appsent's usability, helpfulness, and analytics features, we designed multiple questionnaires. The questionnaires were composed based on the guidelines provided by Ghazanfari et al. [39], and evaluated to answer our research questions in Section II. Questions in the questionnaires were designed using Likert scale measures comprising 1 to 5, where 1 stands for strong disagreement, and 5 stands for strong agreement. The questionnaires were categorized into three sections as shown in Tables III, IV and V, respectively. Having used the questionnaire to gather data to evaluate Appsent, we performed statistical analysis to evaluate the tool's utility based on the information gathered from the questionnaire [40].

Table III. Appsent's Usability Questionnaire

ID	Question
Q1	I think that I would like to use this system frequently.
Q2	I thought the system was easy to use.
Q3	I think that I would need the support of a technical person to be able to use this system.
Q4	I found the system very cumbersome to use.
Q5	I felt very confident using the system.
Q6	I needed to learn a lot of things before I could get going with this system.

Table IV. Appsent's Features' Helpfulness Questionnaire

ID	Question
Q7	The system provides me with multiple levels of information (fine-grained and aggregate information) about users' reviews.
Q8	Using the system to analyze users' review requires less cognitive load for interpretation and is easy to understand than analyzing natural language text (i.e., reviews) manually.
Q9	The visualizations that are used to represent results from review analytics (e.g., Bar graph, Line graph, and Pie graph) are appropriate and easy to understand.
Q10	The system provides information in a timely fashion.

Table V. Appsent's Advanced Analytics Feature Evaluation Questionnaire

ID	Question
Q11	Studying the fine-grained sentiments (positive, negative and neutral) for the top features helped me to understand features that are received well and those that need improvements.
Q12	Studying the fine-grained emotions (happy, sad, anger, fear, disgust and surprise) for top features helped me to understand features that are received well and those that need improvements.
Q13	Timeline analysis on users' emotion, sentiment and star rating help me to understand the changes in users' sentiments and emotions over time.
Q14	Studying the relationships (co-occurring terms) between features and associated issues helps with understanding interconnections between features and issues.
Q15	Overall, the features provided by the system help me to identify improvements to be made to my app/products in order to satisfy users' need.

We advertised for potential participants (app developers) for evaluating Appsent, and 15 respondents agreed to evaluate the tool. The participants were all aware and had previously interfaced with review portals, mobile devices and apps. In addition, all of the participants used mobile devices daily, and were required to report on their prior experience analyzing customer reviews. This allowed us to assess the feedback of those with experience (9 participants) versus those that possessed less knowledge (6 participants). Before performing the evaluation participants watched a brief video tutorial that introduced the basic functionalities of Appsent. This video tutorial acted as a user-manual for participants, and is available online . For the actual evaluation, participants were asked to use Appsent to analyze app reviews (provided by the company above). Appsent was installed on a LG Nexus 5 smartphone with 2GB RAM, a resolution of 1080x1920 pixels and running on Android version 4.4.4 (KitKat). After participants had finished using Appsent, they were asked to answer the three questionnaires above (refer Tables III, IV and V). The scores submitted by each participant were then recorded for analysis.

5 Tool Overview and Evaluation Outcomes

In this section, we overview Appsent's interfaces and main functionalities in more detailed, following by an evaluation of the tool.

5.1 Appsent Interface

Overall, Appsent provides five major functionalities: 1) overview summary for apps, 2) star ranking analysis, 3) sentiment ranking analysis, 4) emotion analysis, and 5) a tutorial feature. These features are included on the home screen of the app. We have not included the home screen Figure given space limitation, however, the video link below[1] provides further details. The left interface in Figure 2 highlights the results of sentiment analysis produced by Appsent. For the sample of reviews above (i.e., 52,705 reviews), 34.3% were classified as positive, 35.5% were classified as neutral, and 30.2% were classified as negative. The sentiments are also visualized based on a timeline scale on the right interface in Figure 2.

The left interface in Figure 3 highlights the results of star ratings, with tabs across the top of the screen for each star rating (1 star to 5 star rating). Below the interface shows the top 10 features that attracted the specific star rating. For example, in Figure 3 it is shown that a specific feature attracted 1-star rating 182 times. Similar to the interface on the right in Figure 2 for sentiment analysis, the star ratings may also be analyzed on a timeline scale (refer to the right interface in Figure 3).

---

[1]https://youtu.be/FPI3sBgOxBY

Figure 2: Appsent's sentiment analysis

Figure 3: Appsent's star rating analysis

By clicking on the negative section (in red) of the pie chart in Figure 2, a different view is then shown (left interface in Figure 4). Here the top 10 negatively-rated features are presented (as for stars in Figure 3). However, when the co-occurring button is clicked, the defects or issues associated with the particular feature of interest are retrieved and displayed (refer to the right interface in Figure 4). So for instance, in Figure 4 the top defects associated with the feature butterfly are displayed. There is also an option provided on the Appsent's screen for the end-user to select a specific feature for further analysis in the form of a drop down list. The co-occurring analysis comprises feature-issue (or defect) and feature-feature relationships. For instance, if "facebook" in Figure 4 is selected, the top defects related to this feature will then be shown in the app. In the screen evident on the right in Figure 4 butterfly was associated with "play", "spend", "update", "keep", and so on, as issues. Selecting "Features" at the top of the right interface in Figure 4 will also allow a user to visualize all of the features that were connected to the specific feature in question.

Figure 4: Appsent's sentiment and co-occurrence analysis

Users may go beyond sentiments and examine the correlation between sentiments and emotions for specific features (sentiment and emotion ranking screens are shown in Figure 5). For example, in Figure 5, 819 reviews expressed positive sentiments about the butterfly feature (on the left), while 893 demonstrated users' happiness (on the right).

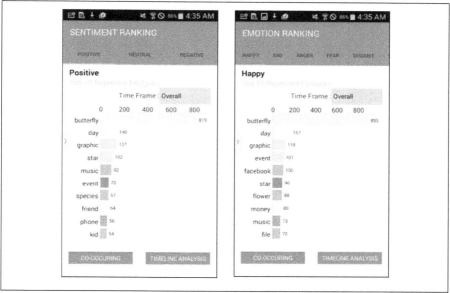

Figure 5: Appsent's sentiment-emotion (multidimensional) analysis

5.2   Appsent Evaluation

We provide the evaluation outcomes for Appsent in this section. The evaluation covers the following three aspects: 1) usability, 2) helpfulness and 3) analytics features.

5.3   Appsent's usability scores

To evaluate participants' perception of Appsent's usability we computed the scores returned from the completed system usability scale

(SUS) questionnaire in Table III. Here, six questions were asked, with respondent scores summed for each question. We stored the scores received from 0 to 4, matching the 1 to 5 Likert scales value selected before converting these to an overall percentage value. This approach was also followed in the analysis conducted in the remaining subsections below. Figure 6 provides a summary of Appsent's usability scores obtained from the experienced (Exp.), non-experienced (Non-Exp.), and overall participants (refer to Section IV for details about participants). Out of six questions, for overall participants, question 4 showed the highest score (Mean = 3.4, Median = 4.0, Standard Deviation = 1.0). The overall lowest score was seen for question 2 (Mean = 2.6, Median = 3.0, Standard Deviation = 0.9). However, for the experienced group, the lowest score was seen for question 1 (Mean= 2.4, Median = 2.0, Standard Deviation = 1.1).

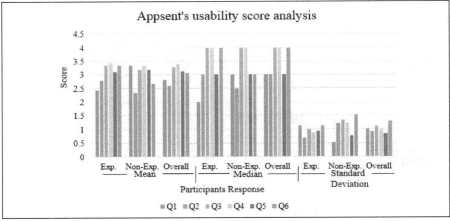

Figure 6: Appsent's usability score analysis

Figure 7 provides the overall usability scores for Appsent (usability assessment questions are shown in Table III). Appsent scored 76.1 for usability (Median = 79.2, Standard Deviation = 17.5). The group-based analysis indicates that the average score of participants that are experienced (Mean = 76.9) was slightly higher than for those who were non-experienced (Mean = 75.0). However, we have observed the opposite result in the median, where the median of participants with no experience (Median = 81.3) showed higher score than the participants with experience (Median = 75.0).

5.4   Appsent's helpfulness scores

To evaluate the helpfulness of the analytics features which are provided by Appsent, we asked participants to answer four questions (refer Table IV). Similar to the usability assessment, we have carried out analyses based on the experienced (Exp.) and non-experienced (Non-Exp.) categories of participants. Figure 8 visualizes these results, demonstrating that question 3 recorded the highest score (Mean = 3.5, Median = 4.0, Standard Deviation = 0.8). The lowest score was seen for questions 1, 2 and 4 in which the values were equal (Mean = 3.4, Median = 4.0, Standard Deviation = 0.7 0.8). Overall, the score was very close to the maximum score (max = 4). From the group based analysis, we observed that the experienced participants' scores for all questions were higher than that of the non-experienced participants. The highest score in the experienced group is for question 2 (Mean = 3.8, Median = 4.0, Standard Deviation = 0.4), however, in the non-experienced group, the score for question 2 was the lowest (Mean = 2.8, Median = 3.0, Standard Deviation = 1.2). We have also observed a reversed pattern for the scores for question 1.

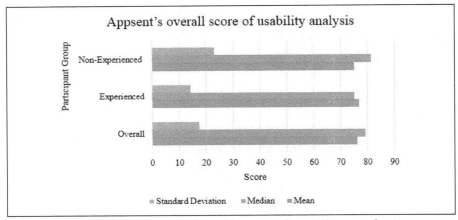

Figure 7: Appsent's overall score of usability analysis

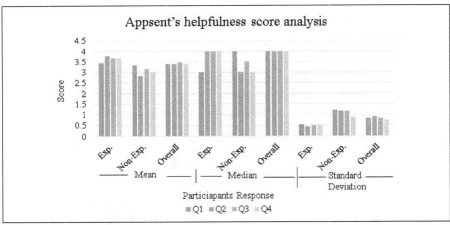

Figure 8: Appsent's helpfulness score analysis

Figure 9 presents Appsent's overall helpfulness scores, with an average of 85.4 returned by respondents (Median = 93.8, Standard Deviation = 17.1). The group-based measures indicate that experienced participants (Mean = 91.0, Median = 93.8, Standard Deviation = 9.4) ranked Appsent higher than those that were less experienced (Mean = 77.1, Median = 81.3, Standard Deviation = 23.3).

5.5 Appsent's analytics scores

To evaluate the analytics features of Appsent, we developed five questions in Section IV (refer Table V). Figure 10 provides a summary of participants' scores in response to these questions. Overall, out of the five questions, questions 3 and 4 show the highest average score (Q3: Mean = 3.5, Median = 4.0, Standard Deviation = 0.7; Q4: Mean = 3.5, Median = 4.0, Standard Deviation = 1.1). Questions 1 and 2 showed the lowest score (Mean = 3.1, Median = 3.0, Standard Deviation = 0.7 0.8). However, when comparing the means these scores are not remarkably different (the mean difference is 0.3). The experienced group (Exp.) assigned the highest score for question 4 (Mean = 3.9, Median = 4.0, Standard Deviation = 0.3) and the lowest score for question 2 (Mean = 3.3, Median = 3.0, Standard Deviation = 0.7). The non-experienced (Non-Exp.) group assigned the highest score for question 3 (Mean = 3.0, Median = 3.0, Standard = 0.9) and lowest score for question 1 (Mean = 2.5, Median = 3.0, Standard Deviation = 0.8).

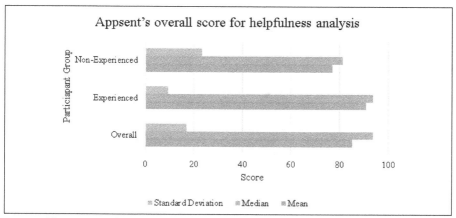

Figure 9: Appsent's overall score for helpfulness analysis

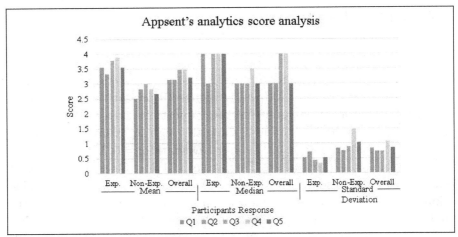

Figure 10: Appsent's analytics score analysis

Figure 11 provides the overall analytics score of Appsent, where it is shown that Appsent scored 82.0 on average for its analytics (Median = 85.0, Standard Deviation = 17.6). In addition, the median score returned for participants with experience (Median = 93.8) was higher than the score for participants with no experience (Median = 81.3).

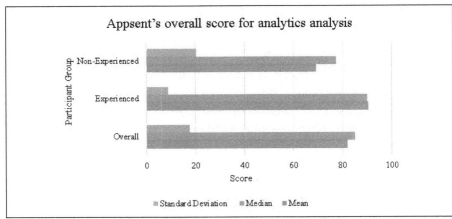

Figure 11: Appsent's overall score for analytics analysis

5.6 Appsent's overall scores

Figure 12 presents a comparison of scores that reflected Appsent's usability, helpfulness and analytics features. The scores of all the questionnaires were aggregated and then visualized. As shown in Figure 12, it is noted that Appsent's helpfulness was rated highest, with the analytics also being scored well by those that evaluated the app. Furthermore, while usability measures were the lowest, these score were also acceptable. Outcomes here suggest that Appsent could be of utility to those handling software reviews logged on app stores. We consider this issue at length in the following section.

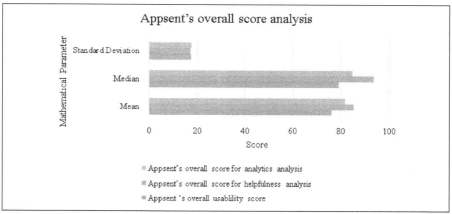

Figure 12: Appsent's overall score analysis

6 Discussion And Implications

We discuss our findings in answering our research questions in this section. Firstly, we discuss RQ1 where we examine Appsent's usability and its utility for analyzing reviews. We next discuss RQ2 and the helpfulness of Appsent. Finally, discussions for RQ3 are provided, where we review the actual analytics support that Appsent provides. *RQ1. How can we design and develop a mobile app for analyzing reviews that is deemed highly usable?*
Results in the previous section show that the average score returned

for Appsent's usability from the participants' evaluations was 76.1 (out of 100). This score shows that although Appsent is only developed as a prototype, it perceived to have good usability. That said, feedback also points to avenues for improving the tool. Overall, users that were aware of other analytics tools did not score Appsent to be much more usable than those that had encountered such a tool for the first time. In fact, the most substantial difference was observed in question one that asked: "I think that I would like to use this system frequently". For this particular question the experienced group had a lower score (2.4 out of 4) than non-experienced group (3.3 out of 4). We received additional feedback from the experienced group indicating that there was no requirement for real-time monitoring of users' opinion from the app-reviews, and hence they believed that there would be no need to use the tool frequently. In fact, the experienced group believed that Appsent is a good tool that they would be happy to use for analyzing bulk reviews periodically. The good feedback received for Appsent was perhaps directly related to the interface design rules that were used during its development (refer Table II). We deliberately aimed to reduce end-user actions and provided end-users with the information that was best suited to their needs [39]. We also ensured the provision of informative responses, memory optimization, eliminated unwanted details, quick loading of the application, and the provision of highly immersive and appealing screens. These principles ensured that Appsent was assessed as usable and the tool was well received.

On the other hand, a few users suggested that the usability must be improved for the app to be very useful for them. For example, a non-experienced software tester who has used the app daily provided the following comment "Usability need (sic) to be improved". Another experienced software developer noted the following: "Poor usability, check standards designs from google". Another user recommended that the app should include a "single page summary" to make the results more readable. We intend to follow these recommendations and improve the app's usability in the future. We discuss the Appsent's usability limitations in Section VII. *RQ2. Does the newly developed app provide functionalities that are helpful?*

Our goal was to design an app that was effective in terms of the structure of the analytics that is provided, the data visualizations, the speed of the analytics process, and the cognitive load that was demanded by the end-users. For the structure of the analytics, we have observed that, in general, the evaluators found Appsent's analytics to be well-structured. The score returned was 3.4 (out of 4). Both experienced and non-experienced participants rated Appsent very similar for this dimension (experienced = 3.4, non-experienced = 3.3). For the cognitive load analysis, the average score was 3.4. Again, this score was nearly close to the maximum possible score of 4.0, which indicates that little cognitive load is required to understand and utilize Appsent. That said, we must also consider the video tutorial that was played before participants used the tool, as this may have enhanced evaluators' perception of Appsent. In addition, experienced evaluators scored Appsent much higher than those that lacked experience using analytics tools (experienced = 3.8, non-experienced = 2.8). This suggests that background knowledge of data analytics influenced evaluators' scores for Appsent positively. Since the experienced group had the understanding of the limitations of the data analysis, their expectations of Appsent's analytics results could have been less than the non-experienced group. Thus, reviewing helpful analytics features addressing the limitations of other tools may have resulted in the higher scores from this group.

Users were mostly happy with the analytics feature of the app. One user described the analytics part of the app as "really good" and "provides fast summary allow to look at different levels". Another user explains that the analytics is "very useful" and "give(s) a clear understanding". Another user mentioned that the app provides a "very good analytic approach. Help(s) to react fast" to users' reviews.

The data visualization score was found to be also high (3.5 out of 4). This score indicates that Appsent presented data visualization results adequately, which helped the participants to understand the opinions of end-users present in the app-reviews. Looking at the average score of the experienced and non-experienced groups of evaluators, we ob-

221

served that the experienced group had a higher score (3.7) than the non-experienced group (3.2). This indicates that having prior knowledge of data visualization enabled the experienced group to understand Appsent's results better than the non-experienced group. The score for the speed of analytics was 3.4 out of 4. This score indicates that Appsent provides analytics results in a timely manner. Results were not very different across groups, although the experienced group returned a higher score (3.7) than non-experienced group (3.0). This suggests that, since the experienced group had prior knowledge about the complexity of text analytics, their expectations of the processing speed of Appsent were lower than the non-experienced group. By performing an overall analysis of the scores of the questionnaires that related to RQ2, Appsent application was regarded by the evaluators for its provision of helpful functionalities (Mean = 85.4, Median = 93.8, Standard Deviation of 17.1). Based on this overall score, and notwithstanding the prototypical nature of Appsent, we believe that analytics features of Appsent have been well designed and Appsent's data analytics features provided user-friendly support for the participants in understanding end-users' opinions present in the app-reviews.

RQ3. Does the developed app provide suitable interpretative analytics on multiple levels?

The research question aims to interpret the actual meaning behind the scores submitted by participants while answering the questionnaires that were related to data analytics features of Appsent (refer Table V).We specifically requested that participants evaluate the sentiment analysis, emotion analysis, timeline analysis, and co-occurring analysis features that are provided by Appsent. According to the evaluation results, the sentiment analysis outcomes provided by Appsent was found to be 3.1 out of 4. This score indicated that the sentiment analysis functionality helped the participants to understand the features that were well received by Appsent's end-users, and also those features that needed fixing. Evaluators that were experienced awarded a higher score (3.6) than the non-experienced group (2.5) for the sentiment analysis feature. A similar pattern was observed for the timeline analysis scores, where the experienced group graded

Appsent higher (3.8) than the non-experienced group (3.0). These results indicate that the non-experienced participants had a higher expectation of Appsent sentiment analysis outcomes compared to experienced participants.

Participants' evaluation of Appsent's co-occurring features was 3.5 out of 4. From the group-based analysis, it was found that the average score of the co-occurrence analysis from the experienced group was relatively high (3.9 out of 4), compared to those of the non-experienced group, where the score was found to be 2.8 out of 4. This statistic for the experienced group was very close to the maximum score of 4. In fact, all participants stated that by knowing the defects/issues that were related to features helped them to gain a better understanding of end-users' opinions, and hence co-occurrence analysis in Appsent proved to be a strong data analytics feature. By analyzing the scores of all five questions that were aimed at answering RQ3, the average score was found to be 82 out of 100 (Median = 85.0 and Standard Deviation = 17.6). Based on this result, it may be inferred that Appsent application helped users to understand complex relationships between the entities of interest expressed by end-users. As highlighted in Section II, we have identified the gap where there was no co-occurrence analysis feature in the commercial applications available. This feature is innovative for any sentiment analysis app if users are to understand the relationships among end-users' opinions, features or issues [20]. Thus, we believe that co-occurrence analysis is a robust data analytics approach that can help to extract meaningful insights from app-reviews.

7 Threats to Validity

We concede that there are limitations to the work that is presented here. This project strongly focused on data analytics, and thus Appsent was developed from that perspective as a proof of concept that a mobile app would be of utility as an analytics tool. To this end, we did not examine multiple interface designs before finally settling on a specific design. We acknowledge that the app (in it's current format) might have some usability limitations, as indicated by some users during the evaluation phase. However, improvement of Appsent's usability features is planned for the future releases.

Further, only one set of app-reviews was analyzed as a part of this study, which may not provide a wide range of testing data for Appsent. However, as the data contained over 50,000 reviews, we believe that the data was adequate for the evaluations that were performed. Also, there were only fifteen participants present for the evaluation of Appent, which could potentially affect the external reliability of the study. We also accept that there is a possibility of the presence of miss-spelled features and issues that might have affected the result of the stemming operation in the text pre-processing stage. Lastly, the sentiment analysis and emotion analysis techniques used in this work may suffer from limitations, although these were previously assessed as adequate by numerous experiments. On the likelihood that these techniques may not always perform accurately, there is a possibility that some of the features or issues could have been misclassified. In this study, we have used a smartphone (LG Nexus 5) that has good hardware specifications (at the time that the study was conducted in May 2017). Hence, all the participants reported Appsent's data analytics and user interfacing performance to be fast, consistent and smooth. That said, there is scope to benchmark Appsent's performance on other devices with different hardware specifications.

8   Conclusion

On the premise that there are numerous text analytics apps, but these lack pertinent features, we have designed and developed a prototype that potentially address several gaps in such apps. Our tool, Appsent, runs on mobile devices and analyses end-users' reviews logged about apps. We used text mining techniques and best practice design principles to provide a robust data analysis framework. By means of empirical evaluation, it was found that Appsent was successful in extracting knowledge of interest from app-reviews, and presented this knowledge in an understandable way through meaningful visualization methods. Appsent software may help providers (or developers) to quickly analyze users' feedback to make rapid decisions in terms of product improvements. Appsent is composed of well-structured modules, and generates efficient data analytics results. These results of the analytics can be visualized in meaningful ways. From these results, the product developers may determine which end-user requirements need urgent attention and fixing. A distinguishing feature of Appsent is its provision of functionalities that allow users to examine the relationships (co-occurring terms) between features and associated issues, which helps with understanding interconnections between features and issues. Based on the evidence provided in this paper, we believe that Appsent could be further developed to provide powerful analytics in understanding users' opinions in text reviews belonging to any domain, which could increase the response speed of product developers towards addressing end-users' requirements.

References

[1] Clifford, C. By 2017, the App Market Will Be a 77 Billion Industry. 2017; Available from: https://www.entrepreneur.com/article/236832.

[2] Statista, Number of apps available in leading app stores as of March 2017. 2017.

[3] Statista. Smartphones - Statistics & Facts. 2017; Available from: `https://www.statista.com/topics/840/smartphones/`.

[4] Vasa, R., et al., A preliminary analysis of mobile app user reviews, in Proceedings of the 24th Australian Computer-Human Interaction Conference. 2012, ACM: Melbourne, Australia. p. 241-244.

[5] Ghose, A. and P.G. Ipeirotis, Estimating the Helpfulness and Economic Impact of Product Reviews: Mining Text and Reviewer Characteristics. IEEE Transactions on Knowledge and Data Engineering, 2011. 23(10): p. 1498-1512.

[6] Licorish, S., Lee, C., Savarimuthu, B., Patel, P., & MacDonell, S., They'll Know It When They See It: Analyzing Post-Release Feedback from the Android Community., in 21st Americas Conference on Information Systems. 2015. p. 1-11.

[7] Licorish, S.A., B.T.R. Savarimuthu, and S. Keertipati, Attributes that Predict which Features to Fix: Lessons for App Store Mining, in Proceedings of the 21st International Conference on Evaluation and Assessment in Software Engineering. 2017, ACM: Karlskrona, Sweden. p. 108-117.

[8] Fawareh, H.M.A., S. Jusoh, and W.R.S. Osman. Ambiguity in text mining. in 2008 International Conference on Computer and Communication Engineering. 2008.

[9] Keertipati, S., B.T.R. Savarimuthu, and S.A. Licorish, Approaches for prioritizing feature improvements extracted from app reviews, in Proceedings of the 20th International Conference on Evaluation and Assessment in Software Engineering. 2016, ACM: Limerick, Ireland. p. 1-6.

[10] Popowich, F., Using text mining and natural language processing for health care claims processing. SIGKDD Explor. Newsl., 2005. 7(1): p. 59-66.

[11] Achimugu, P., A. Selamat, and R. Ibrahim. A Clustering Based Technique for Large Scale Prioritization during Requirements Elicitation. 2014. Cham: Springer International Publishing.

[12] Fu, B., et al., Why people hate your app: making sense of user feedback in a mobile app store, in Proceedings of the 19th ACM SIGKDD international conference on Knowledge discovery and data mining. 2013, ACM: Chicago, Illinois, USA. p. 1276-1284.

[13] Iacob, C., R. Harrison, and S. Faily. Online Reviews as First Class Artifacts in Mobile App Development. 2014. Cham: Springer International Publishing.

[14] Harman, W.M.F.S.Y.J.Y.Z.M., A Survey of App Store Analysis for Software Engineering. IEEE Transactions on Software Engineering, 2016. 43(9): p. 817 - 847.

[15] Liu, B., Sentiment Analysis and Subjectivity, in Handbook of Natural Language Processing. 2010.

[16] Romero, C. and S. Ventura, Educational Data Mining: A Review of the State of the Art. IEEE Transactions on Systems, Man, and Cybernetics, Part C (Applications and Reviews), 2010. 40(6): p. 601-618.

[17] Dave, K., S. Lawrence, and D.M. Pennock, Mining the peanut gallery: opinion extraction and semantic classification of product reviews, in Proceedings of the 12th international conference on World Wide Web. 2003, ACM: Budapest, Hungary. p. 519-528.

[18] Aggarwal, C., & Zhai, C., Mining Text Data. 2012: Springer Science Business Media.

[19] Iacob, C. and R. Harrison. Retrieving and analyzing mobile apps feature requests from online reviews. in 2013 10th Working Conference on Mining Software Repositories (MSR). 2013.

[20] Lee, C.W., et al. Augmenting Text Mining Approaches with Social Network Analysis to Understand the Complex Relationships among Users' Requests: A Case Study of the Android Operating System. in 2016 49th Hawaii International Conference on System Sciences (HICSS). 2016.

[21] AppTrace. May 2017; Available from: http://www.apptrace.com.

[22] AppFigures. May 2017; Available from: https://appfigures.com/.

[23] Apptentive. May 2017; Available from: http://www.apptentive.com/.

[24] SensorTower. May 2017; Available from: https://sensortower.com/.

[25] Haddi, E., X. Liu, and Y. Shi, The Role of Text Pre-processing in Sentiment Analysis. Procedia Computer Science, 2013. 17: p. 26-32.

[26] Stopwords. 2017; Available from: http://www.ranks.nl/stopwords.

[27] Christopher, D., Manning., Prabhakar, R., & S.,H., Introduction to Information Retrieval. 1 ed.: Cambridge University Press.

[28] Toutanova, K., et al., Feature-rich part-of-speech tagging with a cyclic dependency network, in Proceedings of the 2003 Conference of the North American Chapter of the Association for Computational Linguistics on Human Language Technology - Volume 1. 2003, Association for Computational Linguistics: Edmonton, Canada. p. 173-180.

[29] . Standford NLP API. 2017; Available from: http://nlp.stanford.edu/.

[30] Richard Socher, A.P., Jean Y. Wu, Jason Chuang, Christopher D. Manning, Andrew Y. Ng and Christopher Potts. Deeply Moving: Deep Learning for Sentiment Analysis. 2017; Available from: `https://nlp.stanford.edu/sentiment/`

[31] WordNet. 2017; Available from: `https://wordnet.princeton.edu/`

[32] Shneiderman, B., Designing the User Interface: Strategies for Effective Human-Computer Interaction. 1997: Addison-Wesley Longman Publishing Co., Inc. 639.

[33] Android Studio. 2017; Available from: `https://developer.andriod.com/studio`

[34] SQLite Database. 2017; Available from: `https://www.sqlite.org`

[35] Stanford CoreNLP 3.6.0. 2017; Available from: `http://stanfordnlp.github.io/CoreNLP`

[36] Synesketch 2.0. 2017; Available from: `http://krcadinac.com/synesketch/#download`

[37] MPAndriodChart. 2017; Available from: `https://github.com/PhilJay/MPAndroidChart`

[38] Runaway 2017; Available from: `http://www.runawayplay.com/`

[39] Ghazanfari, M., M. Jafari, and S. Rouhani, A tool to evaluate the business intelligence of enterprise systems. Scientia Iranica, 2011. 18(6): p. 1579-1590.

[40] Faraway, J.J., Practical Regression and ANOVA using R. 2002.

Reasoning about Human Values in Plan Selection for BDI Agents

Stephen Cranefield
University of Otago, Dunedin, New Zealand
stephen.cranefield@otago.ac.nz

Frank Dignum
Umeå University, Umeå, Sweden
frank.dignum@umu.se

Michael Winikoff
Victoria University of Wellington, New Zealand
michael.winikoff@vuw.ac.nz

Virginia Dignum
Umeå University, Umeå, Sweden
virginia.dignum@umu.se

Abstract

There is increasing concern about the ethical implications of artificial intelligence, while at the same time its growing capabilities provide great promise for its deployment in socially beneficial ways. This paper considers the deployment of a particular model for developing proactive goal-directed intelligent software—the BDI agent architecture—in application areas where it is desirable for the agent to generate behaviour that respects and optimises for the user's human values, e.g. social assistive technology. We present a mechanism for mapping an agent's current goal and set of plans to an optimisation objective and set of constraints, that are then solved to guide the selection of plans used by the agent to satisfy the goal. We describe the implementation of this technique for the Jason BDI agent-development platform, and illustrate its use with two example scenarios.

The paper provides an update to a prior report on this work, giving more details, enhancements, a correction, and an additional example application.

1   Introduction

Due to dramatic advances in the capabilities of artificial intelligence (AI) software and its increasing deployment in industrial and government agencies, the ethical development and deployment of AI has become a hot topic [2, 7, 8, 21, 27, 16]. At the same time, the use of software and robots in *assistive technology* to support the elderly and people with disabilities is an increasing area of research and development [17]. This paper responds to these two trends by considering the problem of how a prominent model for developing proactive goal-directed intelligent software—the BDI agent architecture [19]—can be enhanced to be part of a socially aware assistive device, such as a home help robot that can provide interactive assistance to a human user (or partner), while respecting the user's human values.

The BDI architecture is a high-level programming model for building a software 'agent' that can proactively and rationally perform goal-directed behaviour, given a library of plans. A key aspect of this architecture is a late-binding mechanism where one of possibly many plans is selected to satisfy a goal at the time the goal is created by the agent. As each plan may generate new subgoals when it is executed, this mechanism leads to a large space of possible agent behaviours. This can make the agent highly adaptive to changing situations. However, this approach also makes it difficult to predict or constrain the side effects of different solutions for a goal. In contrast, agents that are capable of ethical decision-making should generate behaviour that is informed by principles, values, and social norms. Focusing on values, we consider that the various decisions and actions of the agent may have an impact on the user's values. For some (but not necessarily all) goals, it is necessary to create an overall plan that is optimised to leave the user's values as satisfied, overall, as possible. This is complicated by the fact that values can be in conflict (e.g. satisfing the user's desires and maintaining their sense of healthy living may lead to different food choices). Therefore, we present a multi-criteria optimisation approach to selecting plans to bring about a given goal. We present a translation from a goal and a set of agent plans to an optimisation objective and set of constraints for an optimisation solver. The optimal solution is then used to constrain the BDI agent to follow a set of preselected plans. We describe the implementation of this technique for the Jason BDI agent-development platform, and illustrate its use with two example scenarios.

This paper provides an update to our previous work [5], including a more detailed presentation of the optimisation objective, two new features available when annotating Jason plans with their effects on human values, a correction to the mapping from a goal-plan tree to a set of constraints, a description of a now complete implementation of the approach, and a new example scenario that illustrates how considering human values can lead to varying agent behaviour over time. We also show details of the plans in our two scenarios.

The structure of this paper is as follows. Section 2 gives a brief overview of the two scenarios we use to illustrate our approach. Section 3 discusses the concept of human values that we use in this work, and Section 4 introduces the BDI agent architecture and our BDI agent model of our first scenario. Section 5 presents details of our model of values, our approach of annotating plans with their expected value changes, the translation of annotated plans to an optimisation problem, and the use of the solution to guide the BDI plan execution in Jason. The results of the first scenario are also presented in this section. Section 6 discusses enhancements of our approach to track the user's value-related information over time, and uses a second scenario to illustrate how reasoning about values can generate varying agent behaviour over time for the same goal, due to the changing levels of satisfaction for the user's values. Section 7 concludes the paper with a summary of the approach and suggestions for future work.

2 Example scenarios

We illustrate our approach using two scenarios:

1. The *home-help robot scenario* considers a robot that assists an elderly or disabled person with their everyday activities. We focus on a highly simplified example in which the robot chooses the meal and helps the user to prepare and consume it [14]. In order to decide on the most suitable meal, the robot should consider the user's desires, health, wealth and concern about sustainability.

2. The *email assistant scenario* considers an agent that monitors the user's email inbox for product and service advertisements, classifies them, and takes one of several actions: leaving them unaltered, deleting them, or adding informational flags. We consider the specific case of advertisements for fast food, where the potential flags to be applied indicate that the food should only be eaten occasionally, or is a (relatively) healthy option. When choosing an action, the agent must consider the user's autonomy, desire and health.

3 Human values and value-sensitive design

Given the central role of social media in many people's lives and the significant computational power of modern smartphones, there are many opportunities to develop sophisticated software applications to interact intelligently with users and aid them in their personal and social activities. However, along with the opportunities comes the responsibility for the software to understand, respect and enhance the personal values of the users. This is the focus of the field of *value-sensitive design* (VSD), which is based on the principle that human values should be an explicit consideration during the entire software design process [24, 9]. In this work, we consider the implications of a VSD approach to the BDI agent architecture, and in particular its dynamic selection of plans to achieve goals. In this section, we introduce the model of human values that we adopt.

Social psychologist Shalom Schwartz developed the *values theory*, which considers human values to be personal guiding principles that have the following features: 1) they are strongly connnected with our emotions when reinforced or challenged, 2) they represent desirable goals that motivate our actions, 3) they are generic, transcending specific actions and situations, 4) they serve as criteria when making choices, 5) we have our own personal hierarchy of the importance of different values, which is characteristic of us as individuals, and 6) multiple values may be relevant to any situation, and we must make a trade-off between them when choosing our actions [20]. The theory defines ten basic values, distinguished by their motivation, and proposes that these are "likely to be universal" as they are grounded in "universal requirements of human existence". Figure 1 shows these values in a circle, organised in two dimensions: openness to change vs. conservation, and self-transcendence vs. self-enhancement. Adjacent values have similar motivations and are generally congruous, while values in different quadrants are in conflict.

In the home-help robot scenario, the relevant user values are self-enhancement, conservation and self-transcendence. However, these are very abstract values that are difficult to apply to the specific choices to be made in the scenario. Following [23], we adapt the concept of a *values hierarchy* to specialise abstract values (such as self-enhancement) to concrete problem-specific values (such as following the user's preferences), and then to specific (competing) goals for the agent. This is shown in Figure 2. Note that values and related goals may conflict. If the meal most desired by the user is neither sustainable nor healthy, the robot must choose which value(s) to support, at the cost of others. We address this by using multi-criteria optimisation when selecting a plan for each goal, with the impact of the plans on each value weighted by the current salience of the value.

Figure 1: Schwartz's value model

Figure 1. Theoretical model of relations among ten motivational types of value

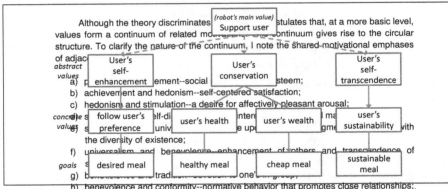

Figure 2: A values hierarchy for the home-help robot scenario

4 BDI agents

The belief-desire-intention (BDI) architecture [19] is a general model for intelligent software ("agents") acting in complex environment, which is based on a theory of human practical reasoning [3]. Definition of the key terms vary, but broadly speaking, the BDI architecture is based on the idea that an agent should generate its behaviour by explicitly reasoning about: (a) its knowledge, which may be uncertain and limited, and is therefore stored as *beliefs*, (b) its *desires*, representing states of affairs that the agent would like to occur, and (c) its *intentions*, which are selected means for bringing about the agent's goals, selected from its desires [6]. The inclusion of intentions is motivated by the agent's potential limitation of resources and the need to focus on one or a few goals at a time. Intentions correspond to currently active *plans* for reaching a desired state of affairs. An important consideration in BDI reasoning is how to control the agent's *commitment* to an intention [3], so that it does not blindly pursue intentions that cannot be fulfilled, swap between alternative intentions in an unproductive manner, or fail to take advantage of new, more fruitful, options to satisfy other desires.

Typically, BDI agents are provided with a library of predefined *plans*, each associated with a goal it will achieve when successful, and a *context condition* that expresses the circumstances in which that plan is applicable. These plans may contain actions and new (sub)goals. Executing a plan can therefore create new goals, and the plans for these goals are selected dynamically at run time. This late binding of plans to goals means that BDI agent programs are highly adaptive, and can generate a very large space of possible behaviours, especially when a failure-handling mechanism is used to allow an alternative plan to be tried when a plan fails [26].

BDI agents operate by following a reasoning cycle. Each iteration of the cycle involves (i) sensing the environment, resulting in a set of percepts, which update the agent's beliefs, (ii) choosing a desire to pursue, (iii) adopting (or retaining) a plan for that desire, and (iv) executing the next step of the plan for that intention.

Plans for the home-help robot scenario

In this work we use the Jason agent platform [1], which implements and extends the AgentSpeak(L) BDI agent programming language [18]. Listing 1 shows an example agent program in Jason. This contains an initial goal and a set of plans for the home help robot

```
!user_eat.                          @make_toast
                                    +!prepare(toast, home) <-
@user_eat                                 toast.
+!user_eat <-
      !choose(Food, Provider);      @order_evil_pizza
      !prepare(Food, Provider);     +!get_pizza(evil_pizza) <-
      !consume(Food).                     ...

@choose_pizza                       @get_local
+!choose(pizza, Provider) <-        +!get_pizza(local_pizza) <-
      !choose_provider(Provider).         ?weather(W);
                                          !pick_up_pizza(W, local_pizza).
@choose_frozen
+!choose(frozen, home).             @collect_local_pizza_by_car.
                                    +!pick_up_pizza(W, local_pizza) <-
@choose_toast                             ...
+!choose(toast, home).
                                    @collect_local_pizza_by_foot.
@choose_evil                        +!pick_up_pizza(W, local_pizza)
+!choose_provider(evil_pizza).          : W = good <-
                                          ...
@choose_local
+!choose_provider(local_pizza).     @eat_pizza
                                    +!consume(pizza) <-
@get_pizza                                eat_pizza.
+!prepare(pizza, Provider) <-
      !get_pizza(Provider).         @eat_frozen
                                    +!consume(frozen) <-
@make_frozen                              eat_frozen.
+!prepare(frozen, home) <-
      microwave.                    @eat_toast
                                    +!consume(toast) <-
                                          eat_toast.
```

Listing 1: A sample of plans for a home help robot

scenario, based on those presented by [5]. We assume that the robot comes from the factory with some generic plans, but can also learn or be told plans by the patient or an authorised person, such as a doctor or relative. We show the plans specific to the goal of helping the patient plan, prepare and consume his/her meal.

The code in the listing begins with an initial goal[1] for the robot: !user_eat, meaning that its goal is for the user to eat. The robot

---

[1]In practice, this goal would be acquired dynamically through observation of or interaction with the patient.

will work together with the patient to satisfy this goal.

The '!' prefix indicates that the initial goal is an *achievement* goal, i.e. it is a goal that the robot needs to proactively bring about. Jason also supports *test* goals, prefixed by '?', representing queries that can be answered by looking up the belief base or by user-provided rules (Horn clauses) that infer information from the belief base (Listing 1 contains no beliefs or rules), and belief addition and deletion goals, prefixed by '+' and '-', respectively.

Following the initial goal, the listing presents a series of plans. Each plan consists of an optional label, beginning with '@', a trigger event, an optional context condition (following ':'), and the plan body (following '<-'). A trigger event of the form '+!goal' represents a new achievement goal being created. Other types of trigger events include the addition or deletion of a belief. The plan body contains a (possibly empty) sequence of actions and goals, and may communicate information back to a higher-level plan by instantiating variables.

The first plan shown is the top-level plan for the !user_eat goal. Its body contains three subgoals: to choose the type of food to eat, to prepare the food, and for the user to consume the food. The remaining plans are options for fulfilling those subgoals and the subgoals created by other plans. For example, the patient could choose to eat pizza, a frozen meal in his/her freezer, or simply make toast. These plans instantiate the **Food** and **Provider** arguments of the !choose goal, and the plan labelled **choose_pizza** creates a subgoal to choose a provider for the pizza. The next two plans are trivial ones with no body, and instantiate the **Provider** variable to an identifier for the multinational Evil Pizza Company or a pizza shop in the neighbourhood.

There are separate plans for preparing pizza, a frozen meal, and toast, and these are followed by several plans related to obtaining the pizza. Home delivery of an Evil pizza can be ordered, or the patient and robot can go to the local pizza company to buy a pizza there. The plan for getting a pizza locally begins with a test goal to check the weather, and the result is passed to the !pick_up_pizza subgoal. There are two plans for that subgoal: one for driving, and

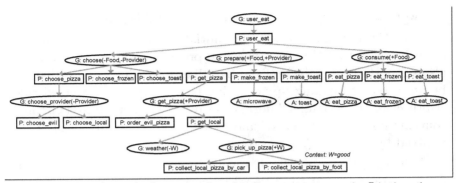

Figure 3: A goal-plan tree for the Jason program in Listing 1

the other for walking, but the `collect_local_pizza_by_foot` plan is only applicable if the weather is good. The details of some plans are suppressed for brevity, with their bodies shown only as ellipses.

The initial `!user_eat` goals and this set of plans can be represented in the form of a *goal-plan tree* [22, 13] as shown in Figure 3. A goal-plan tree is a form of and-or tree that shows the possible ways of satisfying the goal at the root of the tree. Each goal in the tree has zero or more children showing the possible plans for satisfying that goal. Only one of these plans needs to succeed, so a goal is an 'or' node: it succeeds if at least one of its children does. However, a plan can contain multiple subgoals, and they must all succeed. Therefore, a plan is an 'and' node: it succeeds if all its children do. In the figure, we prefix node labels with the node type: G (goal), P (plan) or A (action). For our purposes, we do not need to include test goals and belief addition and deletion goals within goal-plan trees (expect for one exception for test goals, explained in Section 5.3). The figure shows the mode of each argument for a goal: '+' prefixes an argument that is already instantiated when the goal is called, and '−' prefixes one that becomes instantiated when the goal is satisfied. As in Listing 1, Figure 3 suppresses the goals contained within some plans.

238

```
// Beliefs

values([desire,health,wealth,sustainability]).
domain(food, [pizza,frozen,toast]).
domain(provider, [evil_pizza,local_pizza,home]).
random(weather, [prob(good,0.5),prob(bad,0.5)]).

// Plans

@choose_pizza[constraints([(food=pizza)])]
+!choose(pizza, Provider) <-
    !choose_provider(Provider).

@choose_frozen[constraints([(food=frozen),(provider=home)]),
               values([desire(10)])]
+!choose(frozen, home).

@choose_toast[constraints([(food=toast),(provider=home)]),
              values([desire(5)])]
+!choose(toast, home).

@choose_evil[constraints([(provider=evil_pizza)]),
             values([desire(20)])]
+!choose_provider(evil_pizza).

@choose_local[constraints([(provider=local_pizza)]),
              values([desire(20)])]
+!choose_provider(local_pizza).

@get_pizza[constraints([(food=pizza)])]
+!prepare(pizza, Provider) <-
    !get_pizza(Provider).

@make_frozen[constraints([(food=frozen),(provider=home)])]
+!prepare(frozen, home) <-
    microwave[values([wealth(-7), sustainability(15)])].

@make_toast[constraints([(food=toast),(provider=home)])]
+!prepare(toast, home) <-
    toast[values([wealth(-1), sustainability(20)])].

@order_evil_pizza[vbr_leaf,
                  constraints([(provider=evil_pizza)]),
                  values([wealth(-10), sustainability(-5)])]
+!get_pizza(evil_pizza) <-
    ...
```

Listing 2: Annotated plans for the home help robot

243

these constraint annotations would be computed from the plan, but this is left for future work.

- **vbr_leaf** declares that a plan should be treated as a leaf node when performing value-based selection of plans for goals. Any goals that are contained within this plan will have their plans selected using the standard BDI execution mechanism. This increases the efficiency of value-based plan selection when the subgoals of the plan will have no effect on values. In the listing, this is associated with the **order_evil_pizza**—value-based plan selection does not need to consider the various options for ordering pizza (e.g. phone, via the web, or using an app). This is a new feature.

Figure 4 shows the goal-plan tree including all the value annotations. The specific numbers are somewhat arbitrary. The inequalities at the bottom of the figure illustrate the tension between values in this scenario. For example, pizza is the most desired food, but it is unhealthy (but less so for the patient than toast, which has little nutrition). Obtaining pizza from the local supplier is more sustainable (especially if it is picked up by foot), but it is more detrimental for the patient's wealth.

5.3 Mapping annotated plans to a constrained optimisation problem

Given a goal g_0 to be satisfied using value-based reasoning, we translate the annotated goal-plan tree to a program for an optimisation solver. The following translation updates that developed in our previous work [5], which contained some unnecessary elements (input and output variables). It also generated plan choices for subgoals that were not connected to the root goal via the $\prec_n$ relation, sometimes leading to a lack of any feasible solutions.

We now define the translation to a set of constraints for the solver, using the following notation to represent the goal-plan tree. Each node N in the goal-plan tree has a name N^n, an optional constraints annotation N^{ca}, as described in the previous subsection, a type N^t

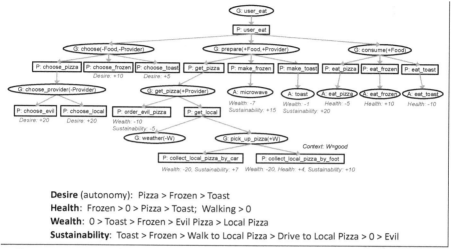

Figure 4: Value-annotated goal-plan tree for the home help robot

$(t \in \{\text{g}, \text{p}, \text{a}\}$ for goal, plan or action), and a list N^C of child nodes: $N^C = \{N_1^C, \cdots, N_k^C\}$ for some k. Leaf nodes also have a values annotation N^{va}, specifying the change to each human value that will result from executing that node (a plan or action)[3].

In the solver program, for each node in the goal-plan tree, we declare a variable with the name of the node, typed to store a vector containing a float for each different human value that is modelled in the agent program. This vector represents the value changes that will be achieved by executing that node. We also declare scalar variables for the domain and random variables defined in the agent's belief base. For each goal node N, we declare a 'choice' vector c_N^n (c stands for "chosen") of size $|N^C|$. The value $c\_N^n[i]$ will be 1 if the ith plan is chosen for goal N, and 0 otherwise.

---

[3]The program that generates the goal-plan tree will create value state change annotations of 0 when some of all of these annotations have not been provided by the programmer.

We then generate the following constraints $c(N)$ for each node N.

$$c(N) = \begin{cases} N^n = N^{va} & \text{if } N^{va} \text{ is present, else:} \\ N^n = \sum_{i=1}^{|N^C|} (N_i^C)^n & \text{if } N^t = p, \text{ else:} \\ \left(\sum_{i=1}^{|N^C|} c\_N^n[i] \right) = 1 & \text{if } N \text{ is the root.} \\ \text{Otherwise (for } M, i \text{ such that } N \prec_i M): \\ c\_M^n[i] = 1 \Rightarrow \left(\sum_{i=1}^{|N^C|} c\_N^n[i] \right) = 1 \\ \wedge \quad c\_M^n[i] = 0 \Rightarrow \left(\sum_{i=1}^{|N^C|} c\_N^n[i] \right) = 0 \\ \wedge \quad \bigwedge_{i=1}^{|N^C|} \left(c\_N^n[i] = 1 \Rightarrow \left(N^n = (N_i^C)^n \right) \wedge (N_i^C)^{ca} \right) \end{cases} \quad (2)$$

The first constraint sets the value-change vector for a node with a values annotation to the numbers in that annotation. The second constraint applies to plans, and states that the value changes resulting from executing a plan is the sum of the value changes that the child goals cause. The third constraint ensures that exactly one element of the choice vector for the root goal node can be set to 1, i.e. only one plan for the root goal can be chosen. The next two lines apply to non-root goal nodes. In this case, there must be a choice vector for the parent goal M. If the plan that contains goal M is chosen, then exactly one of the plans for N must be chosen. Otherwise, none of the plans for N should be chosen, as the goal will not arise. The final line states that whenever a plan is chosen for a goal, the value-changes for the goal are precisely those caused by the plan. Furthermore, any constraint annotation associated with the chosen plan must be applied.

Note that this translation assumes that annotations will only appear on leaf nodes of the goal-plan tree. This restriction can be easily removed in future work: a node's value change annotations can be added to the overall effects of its children.

In some cases, the choice of plans may depend on the value of a variable that is unknown at the time the value-based plan selection is made. For example, suppose that the patient prefers to know all meal choices at the beginning of the day. However, the choices available for collecting a pizza from the local provider depend on the weather. If the weather is uncertain, it cannot be known in the morning whether the more sustainable option of walking to collect a pizza from the local provider will be possible. We handle cases like this in a special way. The program can include facts declaring one or more random variables, together with a list of possible bindings and their estimated probabilities (see the declaration of `weather` in line 4 of Listing 2). Plans may have an initial test goal to look up the value of a random variable[4]. A plan of this form is optimised using an expected value approach. A renamed copy of the subtree rooted at the plan node (omitting the test goal) is made for each possible binding for the random variable. Each copy has a constraint annotation added setting the random variable to a different binding. The value change vector predicted for the plan is then expected value of the plan copies, each weighted by the probability of its random variable binding.

5.4 Satisfying goals using value-based reasoning

Our value-based plan selection process is run dynamically at run-time. If the programmer wishes value-based-reasoning to be used for an achievement goal $!g$, he/she calls the goal $!\text{vbr}(g)$. A supplied plan for **vbr** goals performs the following steps:

1. The plans in the Jason program are analysed to generate a Jason term representing the goal-plan tree for g. This is done by calling a test goal that triggers a set of Jason rules. These retrieve the applicable plans for each goal in the goal-plan tree, and then repeat this recursively for each plan's subgoals.

2. The goal-plan tree is translated to a program for an optimisation solver to satisfy the optimisation objective 1 in Section 5. The translation is done by an external SWI Prolog program [25], which is called using the JPL API[5] via a custom Jason inter-

---

[4]We do not currently handle more complex uses of test goals.
[5]http://jpl7.org

nal action.[6]. The resulting constraint program is then written to a file. We generate constraints in the MATLAB language, using the YALMIP library [11], and we choose the Mosek solver [15]. YALMIP supports the implications in Equation 2 via its `implies` function. This uses a Big-M reformulation of the problem [12].

3. A Jason custom internal action is used to request MATLAB R2019a to load and run the constraint program, via the MATLAB Engine API. This requires a local installation of MATLAB. An alternative option that we have not tested is the use of GNU Octave, which is "largely compatible with MATLAB"[7] and compatible with YALMIP. The program is defined as a function that returns a MATLAB 'structure' mapping each goal to its chosen plan index (if it is part of a chosen plan). This is returned to Jason as a Java map.

4. A Jason *metainterpreter* [4] is used to call the goal g_0, ensuring that the pre-selected plan choices are followed. A metainterpreter is an interpreter for a programming language that is written in the same language. It provides a simple way to extend the features of a programming language. Standard features of the language can be handled by the metainterpreter directly calling built-in constructs of the language (e.g. performing actions), while new ones can be handled by the metainterpreter code. This is an alternative to the approach of dynamically rewriting plans suggested previously [5]. The metainterpreter enables the plan choices to be provided when calling a goal. When the goal or any of its subgoals are being evaluated, instead of choosing the first plan that matches the goal (as in standard Jason BDI execution), the metainterpreter retrieves the list of all relevant plans for the goal, and chooses the one

---

[6]It would be possible to do this translation using Jason rules, but for rapid prototyping, we found SWI Prolog with its richer standard library to be more convenient.

[7]https://www.gnu.org/software/octave

at the preselected index in the list. The metainterpreter is then applied recursively to the goals in the selected plan.

5.5 Results

The internal representation of the problem has 184 variables and 402 constraints.

In a situation where all four values (desire, health, wealth and sustainability) have equal salience, their targets are all 100, and their value states are $(110, 50, 80, 20)$, the best choice (found by the optimisation solver in an average of 0.6794 seconds[8]) is to eat a frozen meal, with value state change $(10, 10, -7, 15)$[9]. An initial value state of $(20, 80, 90, 80)$ leads to ordering a pizza from Evil Pizza and a value state change of $(20, -5, -10, -5)$. A value state of $(10, 70, 100, 80)$ results in ordering pizza from the local provider, giving an expected value state change of $(20, -3, -20, 8.5)$. If wealth is the main concern, e.g. given a value state of $(100, 100, 0, 100)$, then toast is chosen, with a value state change of $(5, -10, -1, 20)$.

6 Balancing values over time

In Section 3, we noted that the state of a particular value (i.e. its level of satisfaction) will generally decay over time, and that the salience of a particular value to a goal may also vary. These changes can then cause the selected plans for a given goal to change over time. In this section we illustrate this with our second scenario: a user assistant agent that (amongst other things) filters and flags advertisements that arrive by email. For illustrative purposes, we

---

[8]This was calculated by summing YALMIP's *yalmiptime* and *solvertime* variables over 10 runs under Windows 10 on a 1.6GHz Intel Core i5-8365U CPU. We currently do not specify bounds for all variables, which YALMIP warns will lead to a poor Big-M relaxation. Providing these bounds is technically simple, but requires domain-specific knowledge.

[9]These results differ from the ones reported earlier [5] due to the corrections made to the generated constraints in this paper.

```
// Beliefs and rules

values([autonomy,desire,health]).
domain(product, [pizza,burger,curry]).
domain(stars, ['*','**','***']).

eval(Value, Func, [product, stars], Multiplier*Amount) :-
    top_goal_bindings(B) &
    .member(product=P, B) &
    .member(stars=S, B) &
    base_amount(Value, P, S, Amount) &
    multiplier(Func, Multiplier).

base_amount(health, pizza, '*', -10).
base_amount(health, pizza, '**', -5).
base_amount(health, pizza, '***', 0).
base_amount(health, burger, '*', -6).
base_amount(health, burger, '**', -3).
base_amount(health, burger, '***', 0).
base_amount(health, curry, '*', -12).
base_amount(health, curry, '**', -6).
base_amount(health, curry, '***', 0).

base_amount(desire, pizza, '*', 15).
base_amount(desire, pizza, '**', 10).
base_amount(desire, pizza, '***', 5).
base_amount(desire, burger, '*', 10).
base_amount(desire, burger, '**', 7.5).
base_amount(desire, burger, '***', 5).
base_amount(desire, curry, '*', 5).
base_amount(desire, curry, '**', 3).
base_amount(desire, curry, '***', 1).

multiplier(fn_keep, 0.5).
multiplier(fn_flag_occasional, 0.4).
multiplier(fn_flag_healthy, 0.6).

// Plans

@keep[values([health(fn_keep(product,stars)),
              desire(fn_keep(product,stars))])]
+!handle_email_ad(MsgID, Product, Stars). // Keep email - nothing to be done

@delete
+!handle_email_ad(MsgID, Product, Stars) <-
    delete(MsgID)[values([autonomy(-5)])].

@flag_occasional[constraints([[(stars='*')]])]
+!handle_email_ad(MsgID, Product, Stars) <-
    flag(MsgID, occasional)
    [values([health(fn_flag_occasional(product,stars)),
             desire(fn_flag_occasional(product,stars)),
             autonomy(-1)])].

@flag_healthy[constraints([[(stars='***')]])]
+!handle_email_ad(MsgID, Product, Stars) <-
    flag(MsgID, healthy)
    [values([health(fn_flag_healthy(product,stars)),
             desire(fn_flag_healthy(product,stars)),
             autonomy(-1)])].
```

focus on a simple version of this scenario. We suppose that a classifier is applied to all arriving email messages, and those that appear to be advertisements cause the BDI agent to receive a new goal !handle_email_ad(MsgID, Product, Stars), where the arguments are the message identifier, the type of product, and a health star rating from one to three stars. Listing 3 shows an excerpt of the Jason program for this agent. It begins with beliefs listing the relevant values (autonomy, desire and health) and the variables product and stars together with their domains. For the sake of brevity, this agent only handles advertisements for pizza, burgers or curries. The program ends with the plans for the handle_email_ad goal. There are four options: do nothing, and therefore keep the email in the user's mailbox, delete the email, add a flag to the email indicating that the food is one that should only be eaten occasionally, or flag it as advertising a healthy option. We assume that these plans will be selected only via value-based reasoning, so they do not include context conditions, as would be used in standard BDI plan selection. Instead, constraint annotations are provided for use in value-based reasoning.

The value annotations in these plans illustrate a new feature that our value-based reasoning mechanism supports: value-changes that are functions of one or more variables. For example, the first plan states that the change to health resulting from this plan should be computed by the function fn_keep applied to the specific product and star rating for the advertisement. Lines 7 to 27 of the listing show how this and the functions used in other plans are evaluated. This is domain-specific and could be implemented in any way the programmer likes. In the listing, the eval predicate is defined to look up the bindings of the product and stars in the original goal (which are saved as beliefs when the goal is passed to the vbr plan). A look-up table base_amount is defined to give a value-change for the given combination of product and star rating, and then a multiplier is applied, with the value of the multiplier depending on the name of the function (fn_keep, etc.).

Omitted from the listing are new features to handle the passing of time between invocations of the handle_email_ad. In particular,

251

the value state is stored as a belief that includes the latest modification time. When a new goal is called using value-based reasoning, a user-defined decay function is called for each value, giving an updated value state. Different values may decay at different rates; we choose somewhat arbitrary values of 0.5 per time unit for autonomy, 1 for desire and 0.25 for health. The decayed value state is used as $S(v)$ in the optimisation objective 1. The salience $sal(v, g_0)$ is also subject to change over time, in a domain-dependent way. Our BDI agent program for the advertisement-handling scenario uses a default salience of 1 for each value, but then applies an adjustment that depends (in general) on the time and the `product` and `stars` values associated with the goal. We choose to adjust the salience of desire by adding 1 in the time periods associated with lunch and dinner (with the baseline salience for all values set to 1).

Suppose the user receives email advertisements for one-star (unhealthy) pizza every half hour from 9am to 12 noon, the target value vector is $(100, 100, 100)$ for autonomy, desire and health, in that order, and the value state vector at 9am is $(68, 79, 40)$. Simulating the resulting series of `handle_email_ad` goals, we find that the agent alternates between deleting and flagging advertisements (as occasional choices) for the first four advertisements. The fifth and sixth advertisements are deleted, and the final one is flagged. Thus, we can see that the agent's behaviour is dynamic as it balances the effects on values over time.

7 Conclusion

In this paper, we have presented an approach that enables a BDI agent to respect the values of a human user when choosing its plans. Our approach involves annotating plans with their predicted effects on the user's values, and performing a multi-objective optimisation to choose solutions for goals for which the programmer has chosen the use of value-based reasoning. We presented two scenarios to illustrate the ability of the approach to trade off values in different subgoals of a plan, and to generate time-varying behaviour as the user's levels of satisfaction of their values fluctuate. We showed the plans used in our Jason implementations of these scenarios and described how our approach changes the standard BDI execution of Jason by translating a goal-plan tree representation of a goal and the available plans into MATLAB code to solve the optimisation problem, invoking MATLAB to make the optimal plan choices, and then using a Jason metainterpreter to enforce those choices when satisfying the goal.

Future work is needed to reduce the burden of annotating the agent plans, and in particular, generating constraint annotations from the plan's context conditions. The approach is reliant on the agent having accurate beliefs about the user's value targets, satisfaction levels and salience for different goals, as well as the effects of plans on values. Therefore, research is needed on learning or acquiring that information from the user. As value-based reasoning complicates the behaviour of the agent, it would also be beneficial to develop a mechanism allowing the agent to explain its decisions in terms of values.

References

[1] Rafael H. Bordini, Jomi Fred Hübner, and Michael Wooldridge. *Programming multi-agent systems in AgentSpeak using Jason*. Wiley, 2007.

[2] Nick Bostrom and Eliezer Yudkowsky. The ethics of artificial intelligence. In Keith Frankish and William M.Editors Ramsey, editors, *The Cambridge Handbook of Artificial Intelligence*, pages 316–334. Cambridge University Press, 2014.

[3] Michael Bratman. *Intention, plans, and practical reason*. Harvard University Press, 1987.

[4] Stephen Cranefield and Frank Dignum. Incorporating social practices in BDI agent systems. In *Engineering Multi-Agent Systems — 7th International Workshop*, volume 12058 of *Lecture Notes in Computer Science*, pages 109–126. Springer, 2019.

[5] Stephen Cranefield, Michael Winikoff, Virginia Dignum, and Frank Dignum. No pizza for you: Value-based plan selection in BDI agents. In *Proceedings of the Twenty-Sixth International Joint Conference on Artificial Intelligence*, pages 178–184. ijcai.org, 2017.

[6] Frank Dignum, David Kinny, and Liz Sonenberg. From desires, obligations and norms to goals. *Cognitive Science Quarterly*, 2(3-4):407–430, 2002.

[7] Virginia Dignum. Ethics in artificial intelligence: introduction to the special issue. *Ethics and Information Technology*, 20(1):1–3, 2018.

[8] Virginia Dignum. *Responsible Artificial Intelligence - How to Develop and Use AI in a Responsible Way*. Artificial Intelligence: Foundations, Theory, and Algorithms. Springer, 2019.

[9] Batya Friedman, Peter H. Kahn, Alan Borning, and Alina Huldtgren. Value sensitive design and information systems. In Neelke Doorn, Daan Schuurbiers, Ibo van de Poel, and Michael E. Gorman, editors, *Early engagement and new technologies: Opening up the laboratory*, pages 55–95. Springer, Dordrecht, 2013.

[10] Jomi Fred Hübner. Annotations in Jason. https://github.com/jason-lang/jason/blob/master/doc/tech/annotations.adoc, 2016.

[11] J. Löfberg. YALMIP : A toolbox for modeling and optimization in MATLAB. In *Proceedings of the CACSD Conference*, Taipei, Taiwan, 2004.

[12] Johan Löfberg. *Big-M and convex hulls*, 2016.

[13] Brian Logan, John Thangarajah, and Neil Yorke-Smith. Progressing intention progression: A call for a goal-plan tree contest. In *Proceedings of the 16th Conference on Autonomous Agents and Multiagent Systems*, pages 768–772. IFAAMAS, 2017.

[14] Derek McColl and Goldie Nejat. Meal-time with a socially assistive robot and older adults at a long-term care facility. *Journal of Human-Robot Interaction*, 2(1):152–171, 2013.

[15] MOSEK ApS. *The MOSEK optimization toolbox for MATLAB manual. Version 9.2.*, 2020.

[16] OECD AI principles. https://oecd.ai/ai-principles, 2019.

[17] Meeko Mitsuko K Oishi, Ian M Mitchell, and HF Machiel Van der Loos. *Design and use of assistive technology: social, technical, ethical, and economic challenges*. Springer Science & Business Media, 2010.

[18] Anand S. Rao. AgentSpeak(L): BDI agents speak out in a logical computable language. In *Workshop on Modelling Autonomous Agents in a Multi-Agent World*, volume 1038 of *LNAI*, pages 42–55. Springer, 1996.

[19] Anand S. Rao and Michael P. Georgeff. BDI agents: From theory to practice. In *Proceedings of the First International Conference on Multiagent Systems*, pages 312–319. The MIT Press, 1995.

[20] Shalom H. Schwartz. An overview of the Schwartz theory of basic values. *Online Readings in Psychology and Culture*, 2(1), 2012.

[21] Tom Simonite. Tech firms move to put ethical guard rails around AI. *Wired*, May 2018.

[22] John Thangarajah, Lin Padgham, and Michael Winikoff. Detecting & exploiting positive goal interaction in intelligent agents. In *The Second International Joint Conference on Autonomous Agents & Multiagent Systems, AAMAS 2003, July 14-18, 2003, Melbourne, Victoria, Australia, Proceedings*, pages 401–408. ACM, 2003.

[23] Ibo van de Poel. Translating values into design requirements. In Diane P Michelfelder, Natasha McCarthy, and David E. Goldberg, editors, *Philosophy and Engineering: Reflections on Practice, Principles and Process*, pages 253–266. Springer, Dordrecht, 2013.

[24] Jeroen van den Hoven. ICT and value sensitive design. In Philippe Goujon, Sylvian Lavelle, Penny Duquenoy, Kai Kimppa, and Véronique Laurent, editors, *The Information Society: Innovation, Legitimacy, Ethics and Democracy In honor of Professor Jacques Berleur s.j.*, pages 67–72. Springer, Boston, MA, 2007.

[25] Jan Wielemaker, Tom Schrijvers, Markus Triska, and Torbjörn Lager. SWI-Prolog. *Theory and Practice of Logic Programming*, 12(1-2):67–96, 2012.

[26] Michael Winikoff and Stephen Cranefield. On the testability of BDI agent systems. *Journal of Artificial Intelligence Research*, 51:71–131, 2014.

[27] James Zou and Londa Schiebinger. AI can be sexist and racist — it's time to make it fair. *Nature*, 559:324–326, 2018.

]

LETTERS

Dear Martin,

Scientific progress is rarely streamlined and linear (as one may surmise based on lecture of published work alone), but undergoes innovation cycles, sometimes by genuine innovations within a given discipline, but oftentimes by cross-fertilisation from other disciplines as well as technological innovations offering novel analytical opportunities. Inasmuch as this may apply to the scientific process, why would it be any different for researchers?

But if such is the case, where do such stimuli come from in the context of researchers, and more specifically, for graduate students?

While most might suggest the necessity to draw on academics from a wide range of disciplines, when it came to stimuli during my time as your PhD student, I could claim the luxury to refer to you. Drawing on your own background and expertise both in science and the arts, your supervision certainly did not fall short of providing a plethora of topical pointers and disciplinary guidance, paired with the openness to address issues with little concern for disciplinary confinements. Importantly, you did so without sacrificing scrutiny on methodology.

It should thus be of little surprise to find you as an advocate of complex social systems modelling, given that only few techniques are able to capture and absorb the diversity of your ideas and concepts at depth and breadth. Your spirit, broad knowledge and interdisciplinary approach is a unique feature that one will struggle to find united in one person. It certainly serves me as a motivation to step back and (re)view issues from unconventional and sometimes uncomfortable perspectives.

With these thoughts in mind, I would like to express my gratitude for the continuous support you gave me throughout my journey at Otago and beyond, as well as the opportunities you provided me with both on a professional and personal level. It has surely changed and shaped me, and certainly into a very different person than the one who joined you.

In an attempt to reflect on our interactions, it fills me with great pleasure to offer you a discussion of agent-based modelling and institutionalism in this volume that poses questions of the kind we might well have debated in our meetings more than half a decade ago – whether at 1pm in the afternoon, or 1am in the morning.

With deepest gratitude,

Christopher

It is an honor to write about Dr. Martin Purvis's life; he was the supervisor during my Ph.D. in Information Science at Otago University. His inspired thoughts were always a motivation to keep going pursuing my goals. In the low moments of my life during my Ph.D., he was an understanding mentor and guided me to the right way out of my problems, and I am very thankful for that. Always very entertaining thorough talks about movies and cinema or practicing volleyball, he was a fun supervisor that kept an eye as well on the social aspects of Ph.D. life. I consider him a motivator who always carries about the research and how better we can produce it.

Martin has demonstrated to be a leader in the Information Science Department. I first worked with him as a Research Fellow under his supervision at New Zealand Distributed Systems project, carrying out the development of a multiagent platform named OPAL. In this project, we have developed a platform for developing multiagent systems in a FIPA complainant architecture. It was an inspiring work that served to build my skills as a researcher in the distributed artificial intelligence field. He was always very worried about the quality of the work, and he used to say that a Ph.D. is a life's work that would be pursued for many years in the career of a researcher. And he was correct on that assumption since the Ph.D. is the most important work of the young researcher. After Ph.D., the researcher keeps in the Ph.D. research field for an extended period developing related research.

His advice was always very inspiring and careful during the construction of research articles, especially. I would describe him as an openminded supervisor with a strong sense of the research goals. As a conscientious adviser was worried not only with the research but also with the social aspects of the Ph.D. life, helping the students overcome problems not only in the study but also in life. The Ph.D. work is very demanding psychologically, and an understanding advisor is essential to help and guide the student to its research goals. His work serves as an inspiration to me as a researcher at Federal University of Ceará in Brazil. We keep in touch, and he was always open to collaboration in research.

In summary, it was great to work with Martin. I am very thankful for his advice during work and Ph.D. life, which helped me be a successful researcher and lecturer in Brazil. And I am keen on future research collaborations.

From Marcos Antonio de Oliveira

Dear Martin,

I would like to be able to express my deepest gratitude to you as a person, friend, and as an academic mentor. However, I already know the words will not be enough to accomplish that.

You have been always available, always ready to offer support on so many different levels, as a friend, teacher, academic researcher, co-worker, and mentor. You have always provided the highest quality mentorship during my junior research work, PhD studies as well as early at the University of Otago. I am now an academic with with my own students and I have no idea how it was possible for you to achieve all of that while maintaining your warmth, cool, and professional insights. Being an academic is tough, and, being a good academic is awe inspiring. You are such a person.

It is clear through this Festschrift that your research interests have inspired your students, me included. We all look into societal problems and deeper relationship between social systems and technology, on the intersection between computer science, humanity, and life in general. I hope your work will continue through the work of your own students, as well as through the work of students of your students – an academic grandparent relationship.

It was you who taught me about people, about myself, and that the research, innovation and brightest ideas always happen in the space between the brains rather than in the brain itself. It is the group, the people, that make the research possible, and fun.

The opportunities of studying and working with you were truly life-changing for me. In the hindsight, I must have been not the easiest student, yet you have shown me how to work, how to motivate myself as well as inspire others, and how working within the academic brotherhood is all about.

Thanks to your examples, deep interests and quality of work, focus on people and students and focus on advancing everyone's interest, you have influenced not only my research interests and work, but inspired me and influenced my life in general. For that, I am deeply grateful.

Thank you, Mariusz

Dear Martin,

This Festschrift volume honours you. The best way I can think of to do this is to comment on your contributions and your attributes. Obviously, these comments are just one person's perspective. However, I hope that, taken alongside the rest of this volume, the different perspectives might together paint a picture that honours you and your contributions.

My earliest memory of you was watching you present at a conference (from memory, Autonomous Agents, in 2000 at Montreal). I recall you setting up your laptop, and noticing that the clock on your laptop was showing some ridiculous hour (like 3am), being still on New Zealand time. I remember thinking that the jet-lag from New Zealand to Canada must be really bad!

Some years later I interviewed for a position in the department of Information Science. You were, of course, as the Head of Department, on the interview panel. I remember having a long discussion about the department. I very much appreciated not just your willingness to take the time to discuss things, but also your openness.

As I settled into the department and got to know it better, one of the things that I greatly appreciated was the level of collegiality: everyone was there to do what was best for the department as a whole. This is certainly not something to take for granted in academia, and I am very grateful that the department had this culture. Culture of course is complex, but the behaviour of leaders matters a great deal, and I believe that this fantastic departmental culture does owe a great deal to your own attitudes and modelled behaviour.

Later, after I had become Head of Department, I appreciated greatly knowing that I could always talk with you about any particular issues (and I certainly did!), and that you would be invariably supportive and helpful, offering insightful and useful thoughts and suggestions.

Indeed, this is one particular thing that I greatly appreciated, and admired, about you: your ability to bring fresh (and sometimes quite different) perspectives to situations. Whether the topic was running the department, research, or teaching, you could always be counted on to bring new ideas. Ideas that others hadn't thought of, and that were original, insightful, and highly valuable.

One of the things that really has impressed me about your research has been not just its originality, but also its breadth. A typical paper might use agent-based simulation to understand the factors that affected the success of historical trading societies in the 10th-13th centuries. Doing this sort of work, and doing it well, requires familiarity with a considerable body of knowledge. In this era of specialisation, it takes a broad mind and unusual person to do this sort of work.

In addition to a broad intellectual curiosity, I also want to acknowledge your strong sense of ethics and of social justice. Again, you led from the front, quietly doing what was right. Whether that was your dietary choices (vegetarianism), or working on research topics that tackled energy management to help address climate change.

I see you as the perfect example of a "classic" academic: incredibly bright, articulate, with a wide range of knowledge, a wealth of original ideas, and a passion for research and education. You are also generous with your time, and never seem stressed or in a hurry. Perhaps because of that you have been so incredibly prolific. I recall you commenting at one point that you could not staple your CV: it had gotten so long that a standard stapler could not cope with even just the list of your publications and other achievements.

Martin, it has been a pleasure and an honour to have worked with you. You have been a quiet inspiration, setting the tone for the department's collegial and collectivist culture, and leading by example as an outstanding, yet humble, academic. This volume is an incredibly well deserved honour, and I hope that my small contribution helps it to suitably honour you and recognise your work over the many years.

Best wishes!

Michael Winikoff

Martin was my PhD supervisor. Back then around 1993, I did not know what research was about, and it was Martin who guided me into the wonderful world of scientific research. He used to be a Physics professor so I was given a topic on developing an optical thin-film multilayer model for machine learning, a very unusual topic by any standard... I never thought I was capable of doing a PhD, but doing this PhD project with Martin essentially changed my life, and set me up for a research and academic career pathway!! I was so fortunate to have him as my supervisor, as he had such a brilliant and sharp mind. Martin is someone who can strike a conversation with you on just about any topic in science. I think I benefited immensely from his broad knowledge, his brilliant intellect, and his open-mindedness as my supervisor. It was him who opened up my eyes to a vast array of possibilities, and it was him who nurtured and fostered my curiosity, not to be afraid of asking hard questions, and to be resilient when things were not travelling well. When I faced challenges and bewilderment in research, it was him who encouraged me and steered me in the right direction. Martin, to me, is more than just my academic supervisor. He was my mentor and a close friend. I will be always grateful to him for being there for me when I needed his guidance the most!

From Xiaodang Li

Nurturing the "norms" seed

Dear Martin,

First of all, I would like to thank you for inspiring, encouraging and challenging me to work broadly on social computing, and in particular, normative systems. I have cherished the numerous conversations we have had on this subject over the years including our on-going debate on whether a 'personal norm' is really a norm or just a preference. My view still is that it is a norm because one can sanction oneself (i.e., beating oneself up) for violating their own personal norm. A preference doesn't necessarily have the sanction component. You may still disagree, and that can continue to be the conversation we have over the years to come! I would like to take this opportunity to provide an overview of the joint work we have pursued subsequent to my PhD, and those I have conducted with other colleagues on normative systems. While my PhD work mainly focussed on creating simulated models of societies that followed norms, my recent works have focussed on studying norms in real-life societies. In what follows, a summary of work on norms in both these areas is provided. As you may know well, norms are essential for smoother functioning of societies – artificial or real. They enable cooperation and collaboration among agents involved. The first branch of my research focuses on simulating societies to study how norms facilitate cooperation and collaboration among agents, and also studying mechanisms associated with how they are proposed, spread, morphed and removed in societies. One of the works focussed on studying the role of norms in establishing cooperation in hunter-gather societies [1]. Another body of work, with several students we have supervised, examined how social factors (including norms) were influential in the success of historical societies such as Maghribs, Genoese and Julfans [2, 3]. These works have also contributed conceptual and executable models, that can be refined further based on new evidences from historians and anthropologists, and they have an impact on how modern societies can be shaped to achieve desirable behaviour. The second branch of research was on studying norms in real-life societies by analysing data available. One of the early works involved the extent of adoption of green norms, particularly the double-sided printing norm at the department of Information Science [4]. Another body of study involves the study of norms and their impact in the software engineering (SENG) community. One of the works investigated coding convention violation in the SENG community [5]. Another focussed on proposing a taxonomy of communication norms in responses of developers to app reviews (in app reviews). Based on this work, a normative system that produces norm-compliant responses has been proposed [6]. A third body of work has focussed on studying norm compliance ('as-is' vs 'should-be') in the Python community [7]. Our work showed that there is a discrepancy in expected behaviour and actual behaviour. A more recent work focussed on studying to what extent politeness norms are violated in SENG communities such as Stack Overflow (i.e., presence of offensive language), which can make members uncomfortable, and as a result they may leave these communities [8]. One of our current efforts is towards building a system that an rephrase offensive comments into polite ones and thereby creating a more welcoming environment particularly for new comers. In the future, I would like to study normative behaviour in large electronic societies such as multi-player online games, to better understand the norm life-cycle model and then use that information to design norm-aware electronic agents [9]. For example, in the gaming context, a non-player character can learn human norms and use it for its own competitive advantage (e.g. deceive human player at the opportune moment). A positive use of norm-awareness, is in the context of human-robot collaboration, where a robot can use the norm-awareness module to learn about norms and subsequently facilitate believable and trustable conversations and interactions with its human partner. Having learnt the norm, the agent can propagate it to other agents and apply the same norm in other contexts to facilitate inter- and intra-robot collaborations. Another area that is of interest is norm entrepreneurship where new norms are proposed by the agent based on it its observation of the external world. Using rapidly advancing machine-learning techniques an agent can promote new and efficient norms. I hope the above provides a very high-level view of some of the work I have pursued in the area of norms, with significant contributions from colleagues and students. I would like to thank you for nurturing (with Maryam and Stephen) and continuing to do so of what had sprouted from the "norms" seeds. You were instrumental in getting this seed off the ground. This small slice of social reality (norms) has fascinated me for these 18 years I have known you, and continues to do even more today. For that I owe a lot to you!

Thanks also for making the research platform 'green' where I could explore a topic of my own choice. Also, thanks for facilitating a family-like research environment which made it a lot easier for a number of research students (including me) to sprout, grow, blossom and fruit. Your efforts have enabled a research forest to grow and flourish, in many parts of the world beyond New Zealand (e.g., Australia, Brazil, China, Germany, Iran, India, and Pakistan). I have also enjoyed co-supervising many students with you over the years and I would like to continue this in the future. Finally, wishing you the best for the years to come! I am in particular looking forward to our discussions on various topics in the years to come. I also pray to God that the years to follow are as productive and enjoyable as the former ones.

Regards,
Tony

www.ingramcontent.com/pod-product-compliance
Lightning Source LLC
Chambersburg PA
CBHW071106050326
40690CB00008B/1136